Contents

Preface...2
Chapter 0: The Blindness of Behavioural Agnosia............4
Chapter 1: Introduction to ABA and Behavioural Principles...10
Chapter 2: The ABCs of Behaviour..............................23
Chapter 3: How Motivation Operates............................29
Chapter 4: Behavioural Control.................................37
Chapter 5: Verbal Behaviour....................................45
Chapter 6: Verbal Communities..................................51
Chapter 7: Relational Frame Theory............................57
Chapter 8: Putting it into Practice...........................64

Preface

Alright, mate? Name's Ibrahim Kamara, but my friends call me Ibs. I've got myself a BSc in Behavioural Sciences, an MSc in Psychological Sciences, and another MSc in ABA. Yeah, I know, it sounds like I got myself into some serious debt, right? But really, I just love learning about how we humans tick, and how we can help each other tick better. Let's talk about this book. It's about something called behavioural agnosia, which is basically when you don't quite get why you do what you do, or why others do what they do. It might sound a bit scary, but don't worry, we're not here to scare you. We're here to help you understand it, and maybe even do something about it. Now, if you're thinking "ABA? RFT? SDs? What the bloody hell is all this alphabet soup?", don't worry, we'll explain it all in plain English. No jargon, no gobbledygook, no bull. Just simple, practical ideas that you can use in your own life. Whether you're a stressed-out single parent, a student struggling with grades, or just someone who wants to make sense of this crazy world, there's something here for you.

Chapter 0: The Blindness of Behavioural Agnosia

In this chapter, we explore a condition called behavioural agnosia, which affects the way individuals perceive and understand human behaviour. We delve into the challenges faced by those with this condition and how it impacts their daily lives.

Chapter 1: Introduction to ABA and Behavioural Principles

This chapter serves as an introduction to Applied behaviour Analysis (ABA) and its fundamental principles. We provide an overview of ABA as a scientific approach to understanding and changing behaviour. We explore the key concepts and techniques used in ABA and their practical applications.

Chapter 2: The ABCs of Behaviour

Here, we delve into the ABCs of behaviour, namely Antecedents, behaviour, and Consequences. We discuss how these three elements interact to shape and maintain behaviours. Understanding these ABCs is essential for implementing effective behaviour change strategies.

Chapter 3: How Motivation Operates

In this chapter, we examine the concept of motivation and its role in behaviour. We explore different types of motivation and how they influence our actions. Additionally, we discuss strategies for enhancing motivation to promote positive behaviour change.

Chapter 4: Behavioural Control

This chapter focuses on behavioural control and how individuals can gain better control over their own behaviour. We discuss self-monitoring, self-evaluation, and self-reinforcement techniques that empower individuals to modify their behaviour independently.

Chapter 5: Verbal Behaviour

Here, we explore the complexities of verbal behaviour and its importance in human communication. We discuss various aspects of verbal behaviour, including language development, communication deficits, and effective strategies for promoting language skills.

Chapter 6: Verbal Communities

Building upon the previous chapter, we delve into the concept of verbal communities. We examine how social and cultural factors influence language development and communication patterns within different communities. We also discuss the impact of verbal communities on individual behaviour.

Chapter 7: Relational Frame Theory

In this chapter, we introduce Relational Frame Theory (RFT), a psychological framework that explores how humans understand and relate to the world through language. We discuss the key principles of RFT and its applications in behaviour analysis.

Chapter 8: Putting it into Practice

In the final chapter, we focus on applying the knowledge gained throughout the book. We provide practical examples and case studies to illustrate how behaviour analysis principles can be implemented in real-life situations. We also discuss ethical considerations and potential challenges when using behaviour analysis techniques.

Call to Action!

As an author, I'd love to hear your thoughts on my book! If you enjoy reading it, I'd be so grateful if you could take a moment to head over to Amazon and leave a review. Sharing your feedback will help me understand what you liked about the book and what aspects you didn't enjoy as much. This valuable insight will allow me to improve and grow as a writer, making future books even better! Also, leaving a review will have an incredible impact on reaching other potential readers. Reviews play a significant role in Amazon's algorithms, helping the book get more visibility and reach a broader audience. Your review can make a real difference in helping others discover the book and decide whether it's the right read for them. Thank you so much for your support and taking the time to share your thoughts. I truly appreciate it, and it means the world to me as an author. Happy reading and reviewing! ●

Chapter 0: The Blindness of behavioural Agnosia

We like to think that our actions and decisions are the result of our own free will and personal preferences. But what if the truth is much more complex than that? What if the choices we make every day are not solely our own, but rather the product of a lifetime of environmental shaping that we are largely unaware of? This is the concept of behavioural Agnosia – the lack of awareness or recognition of the environmental factors that shape behaviour. Just as people with visual agnosia can see objects but fail to recognize their meaning, those with behavioural Agnosia can experience the world around them, but fail to recognize the impact it has on their behaviour. behavioural Agnosia is all around us, yet most people are largely blind to its effects. We are bombarded by environmental stimuli from the moment we wake up in the morning to the moment we go to bed at night. Every interaction we have, every decision we make, every behaviour we exhibit is influenced by a multitude of environmental factors that we are often completely unaware of.

Consider the following scenarios:
- You grab a sugary snack from the vending machine at work because you feel hungry and need a quick energy boost. In reality, the vending machine is a discriminative stimulus that has been reinforced by previous experiences of getting a tasty treat from it when hungry.

- You pick out a specific outfit to wear because it makes you feel confident and attractive. However, your choice of clothing is also influenced by social norms, cultural expectations, and the people you will be interacting with that day.

- You get into a heated argument with a family member because you have different opinions about a particular issue. However, your beliefs and attitudes towards that issue have been shaped by your past experiences, education, and exposure to different sources of information.

Decisions..decisions..Our preferences have largely been shaped by our past environmental experiences

These are just a few examples of how behavioural Agnosia can affect our daily lives. We often make choices based on habit or impulse without considering the impact of the environment. We fail to recognize the power that the environment has in shaping our behaviour, attitudes, and beliefs. But it doesn't have to be this way. By understanding behavioural Agnosia and the environmental factors that shape our behaviour, we can take control of our choices and make intentional decisions that align with our values and goals. Throughout this book, we will explore the principles of behaviourism and the ways in which they can be used to improve our lives and the world around us.

behavioural Agnosia is not an entirely new concept *per se*. behavioural science refers to our inability to notice the multitude of daily marketing prompts or "nudges" as 'nudge blindness,' behavioural agnosia covers more than just the prompts, but the vastness of behavioural and environmental interactions that we become habituated to. However, the foundations of this concept starts with the foundations Behaviourism - the scientific study of all behaviour, animals and humans.

The cornerstone of behaviourism is the idea that all behaviour can be studied scientifically, devoid of any reference to internal mental states. B.F. Skinner, a prominent figure in this field, suggested that human behaviour is determined by environmental variables rather than by individual's free will. In his perspective, our actions are conditioned responses to environmental stimuli. Skinner's work, centred on operant conditioning, illustrated how behaviour is influenced by the consequences that follow it. He demonstrated that a behaviour that leads to rewarding outcomes is more likely to be repeated, while behaviour followed by negative outcomes is less likely to be repeated. From this viewpoint, our behaviours are shaped by a history of rewards and punishments rather than conscious decision-making. In essence, Skinner proposed that free will is an illusion. Our choices, rather than being freely made, are shaped and controlled by the environment. This idea may seem radical, but it's not entirely without merit. Our everyday behaviours, from the mundane to the significant, can often be traced back to learned responses.

In a similar vein, Jordan Peterson, a Canadian psychologist and professor of psychology, echoes Skinner's assertions to some extent. Peterson argues that we are significantly influenced by both our biology and our cultural surroundings. He suggests that we are complex creatures whose behaviour, thoughts, and attitudes are shaped by a combination of genetics and cultural conditioning. Peterson highlights that the interplay of these two factors is unavoidable and substantial in dictating how we act, think, and feel. For instance, the culture we are brought up in influences our beliefs, values, and behaviours. Similarly, our genetic makeup can predispose us towards certain attitudes, behaviours, and even vulnerabilities to specific mental health issues. Underlying both Skinner and Peterson's views is the proposition that our sense of free will might be less potent than we believe. The choices we make, the behaviours we exhibit, are not necessarily the outcome of a free, conscious deliberation. Rather, they could be seen as the deterministic product of our genetics interwoven with cultural conditioning.

The 'Illusion' of Free Will

Picture this: I offer you an apple and a banana and tell you to pick one. The choice is entirely yours. This might seem like a simple demonstration of free will - the idea that we are the masters of our own destinies, and that every decision we make is wholly our own. The power to choose freely feels deeply personal, right? But, what if I told you that your free will is actually an illusion? That were essentially just bundles of atoms reacting predictably to stimuli, and that our choices could theoretically be predetermined?

Imagine if we could rewind time to the moment you picked between the apple and the banana. You'd assume that if free will exists, you could just as easily switch your choice. But what if, under the same circumstances, you picked the same fruit again? What if I told you I could predict your choice with 100% accuracy, 300 milliseconds before you made it? It sounds far-fetched, but back in the 1980s, a scientist named Benjamin Libet used an electroencephalogram (EEG), to demonstrate that our subconscious minds actually decide on a course of action before our conscious minds become aware of the decision. In other words, by the time we think we're making a decision, it's already been made for us.

Other studies have found similar results, using technologies like functional magnetic resonance imaging (fMRI). They've shown that our brains settle on a decision seconds before we consciously make it. It's only after the fact that we convince ourselves that we're actively making the decision. Just like our hearts, our brains just do their thing; we don't consciously control them. This realisation leads to some hard questions. If our choices aren't truly our own, what about things like morality and accountability? Should people be praised or blamed for their actions if their decisions were not truly made freely?

The influence of genetics and environment also muddy the waters of free will. If you're born into a religious family, it's likely you'll also be religious. Similarly, intelligence, which is largely genetic, shapes the decisions we're able to make. If our genes and upbringing limit our options, can we really say we're making free choices? This logic challenges the idea of free will to its core. Some scientists have even likened belief in free will to belief in religion, neither of which can be reconciled with the laws of physics.

The most drastic evidence against free will is found in cases where physical changes to the brain drastically alter a person's behaviour. There have been cases where brain tumours have transformed respected individuals into criminals. When the tumours were removed, their behaviour returned to normal. If free will truly existed, how could a tumour negate it? Such findings are causing shifts in the legal world. Lawyers have begun using fMRI scans to argue that their clients' criminal actions were caused by brain malfunctions, not free choices. This approach led to the abolition of capital punishment in Illinois.

Some scientists, clinging to the idea of free will, suggest that even if our subconscious decides for us, we still have the power to shape our subconscious. But this argument falls apart when you consider that the desire to shape our subconscious is itself a product of the subconscious. So if free will doesn't exist, what's left? Does morality disappear? Does society collapse? Some people jump to these fatalistic conclusions, deciding that if we can't control our destiny, then we're utterly powerless.

Does that mean our upbringing, our surroundings, and societal norms don't affect our decisions? Not quite. We've known for ages that factors like these significantly shape our decision-making. Think about it - someone who's born into a religious family is more likely to stay religious. Things like genetics also weigh heavily on our choices. For instance, intelligence, a trait believed to be hereditary according to Charles Darwin, can influence our decisions. If you are naturally smart, you're more likely to make "intelligent" choices. So, not everyone gets dealt the same hand in life, and even among those who do, not everyone can play those cards the same way.When viewed through this lens, the concept of free will starts to crumble. Some researchers have concluded that believing in free will is akin to believing in religion. Both concepts, they argue, are at odds with the laws of physics.

This perspective brings us to a deterministic viewpoint - the notion that all events, including human cognition, behaviour, decision, and action, are causally determined by an unbroken chain of prior occurrences. It's like all our decisions are just dominoes in a long chain, with each one causing the next. Does this mean that there's no room for free will in our lives? Not necessarily. Although the concept of free will might seem counterintuitive in light of such scientific findings, there's still a lot of debate around this issue. Some argue for a concept known as compatibilism – the belief that free will and determinism can coexist. It suggests that our actions can be both determined by our past, and yet we can still make free decisions. In other words, we can still have control over our actions even if they are predictable or influenced by past events. Others propose that our consciousness is much more complex and not fully understood yet. As our understanding of the brain and consciousness evolves, it's possible that we may discover new insights into the nature of free will.

So, while the concept of free will may seem straightforward at a first glance, it's a deeply complex and philosophical topic that challenges our understanding of human consciousness and decision-making. For now, the jury is still out on whether we truly possess free will or whether our decisions are merely the result of deterministic processes.To illustrate the illusion of free will, let's dive deeper into how we develop preferences. Every person is born with a set of genes - their genetic endowment - that shapes their body and brain in unique ways. Some people might be born with a genetic makeup that makes them more likely to enjoy certain objects, activities, sensations, people, places or foods etc. For Example, are several genes that could potentially contribute to an individual's preference for or craving of sugar. Although the genetic influences on taste preferences are complex and not entirely understood, some genes have been identified to play a role in sugar cravings.

1. **TAS1R2** and **TAS1R3**: These genes code for the T1R2 and T1R3 proteins, which together form the sweet taste receptor. Variations in these genes may lead to differences in the sensitivity to sweet taste, with some individuals having a heightened sensitivity to sweetness, possibly increasing their preference for sugar.

2. **GNAT3**: This gene encodes gustducin, a taste-signalling protein involved in sweet taste perception. Variants of the GNAT3 gene could impact the way sweet taste is detected and processed, potentially influencing an individual's preference for sweet foods.

3. **FTO**: The FTO (fat mass and obesity-associated) gene has been linked to obesity and appetite regulation. Some studies suggest that individuals with certain variants of the FTO gene may have a higher preference for energy-dense foods, including those high in sugar.

4. **DRD2**: The DRD2 gene codes for the dopamine D2 receptor, which is involved in the brain's reward system. Variations in this gene could influence how the brain processes rewards, potentially affecting an individual's preference for rewarding foods like sugary treats.

A person with such a genetic endowment might have more sweet taste receptors on their tongue, or their brain might release more feel-good chemicals like dopamine when they eat sugary treats. So, from the day they are born, they're already more inclined to enjoy sweets.

Then, think about the environment, which includes everything around us: our home, school, friends, family, and broader culture. The environment introduces us to different kinds of foods and teaches us what's considered good or bad, acceptable or unacceptable. For example, if you grow up in a household where dessert is a normal part of every meal, you'll be exposed to sugary foods frequently. Over time, you might develop a strong preference for them because they're not only pleasing to your taste buds but also associated with feelings of comfort, happiness, or reward.

Culture also plays a crucial role too. Different cultures have different dietary norms and expectations. For example, in some cohorts, high-sugar foods are a staple and widely accepted, while in others, they might be frowned upon or reserved for special occasions only. Depending on where you grow up, your culture could influence your food preferences and habits, reinforcing or discouraging your innate susceptibility for sweets.

So, while our genes might make us more likely to enjoy certain things (like high-sugar foods), it's the environment and culture that often guide how these preferences develop. Behavioural Agnosia says people who chose things because they 'want too' fail to recognise the interplay of genetic predispositions and environmental influences (past and present) in shaping our behaviour and choices.

Science of Behaviour

As a field of study, Behaviourism has been around since the late 1800s. However, it wasn't until the mid-20th century that behaviourism gained mainstream acceptance, thanks in part to the work of influential psychologists such as B.F. Skinner and Ivan Pavlov. Despite its long history and proven effectiveness in a variety of settings, behaviourism is still often misunderstood and even stigmatised. Some critics argue that behaviourism is too focused on controlling behaviour and ignores the complexities of human thought and emotion. Others worry that behaviour modification techniques are unethical or dehumanising. However, these negative misconceptions are largely unfounded. ABA, a specific branch of behaviourism, has been shown to be a highly effective treatment for a wide range of conditions, from autism spectrum disorders to substance abuse. And far from being dehumanising, behaviourism actually emphasises the importance of treating individuals with dignity and respect while helping them achieve their goals. By addressing these misconceptions head-on, we can clear the way for a deeper understanding of behaviourism and its potential to improve our lives. Through this book, we aim to present behaviourism in a way that is accessible, engaging, and applicable to readers from all walks of life.

Whether you're a parent looking to improve your child's behaviour, a healthcare professional seeking new ways to help patients, or simply someone interested in understanding the science behind our actions, this book is for you. Join us on a journey of discovery as we explore the fascinating world of behaviourism and uncover the hidden factors that shape our behaviour.

The consequences of behavioural Agnosia can be severe. When we fail to recognize the environmental factors that shape our behaviour, we may make decisions that are harmful to ourselves or others. For example, we may engage in risky behaviours such as substance abuse or reckless driving, without fully considering the impact these actions have on our health and safety. Likewise, when we fail to recognize the environmental factors that shape the behaviour of others, we may misunderstand or misjudge their actions. This can lead to conflicts in personal and professional relationships, and even to discrimination and prejudice. Furthermore, by remaining behaviourally agnostic, we may miss out on opportunities for personal growth and self-improvement. We may continue to repeat negative patterns of behaviour, without realising that we have the power to change them. But again… it doesn't have to be this way.

By learning to recognize and understand the environmental factors that shape our behaviour, we can take control of our choices and make intentional decisions that align with our values and goals. We can also learn to recognize the factors that shape the behaviour of others, and respond to them with empathy and understanding. Throughout this e-book, we will explore the principles of behaviourism and the ways in which they can help us overcome behavioural Agnosia and achieve our full potential. So join us on this journey of discovery, and let's uncover the hidden factors that shape our behaviour and empower ourselves to live more fulfilling and meaningful lives.

📢*Disclaimer*
behavioural agnosia is simply an analogy. An otherwise unproven theoretical concept used to describe the possible phenomenon hypothesised to be experienced by all humans at some point in their lives. It is not a diagnosis or a cognitive deficit but rather a description of the lack of awareness, identification or recognition of the underlying causes of our daily behaviour. This lack of awareness can be due to many factors, such as habituation, the complexity of the environment, longevity of past operant activity, the influence of social and cultural factors, and the way our ongoing experiences shape our behaviour.

👥People with bad grades may experience behavioural agnosia in the form of not recognizing the antecedents or consequences that lead to their poor academic performance.

👤Those who commit crimes may experience behavioural agnosia as they fail to recognize the factors that trigger their criminal behaviour.

■●Individuals with sudden and unexplained mental health issues may experience behavioural agnosia if they are unable to identify the antecedents that lead to their symptoms.

●■People with varying personalities may experience behavioural agnosia if they are unable to recognize the environmental factors that trigger their different personas.

✔●Stressed single parents who may struggle to manage their own emotions and the behaviour of their children, leading to a lack of awareness or understanding of the reasons behind certain behaviours.

●●Friends, family members, or colleagues who exhibit sudden changes in mood or behaviour, particularly if these changes are uncharacteristic or difficult to understand.

It is important to note that the description of behavioural agnosia is not meant to label or stigmatise anyone, but rather to highlight the importance of understanding the antecedents of our behaviour and how they shape our lives. By understanding the role of the environment and other external factors in our behaviour, we can take action to modify our behaviour and achieve our goals.

In essence, behavioural agnosia is a call to action to become more aware of the factors that shape our behaviour and to take control of our lives. It is an opportunity to reduce the total personalization of behaviour due to pure free will and to become more conscious of the role that external factors play in shaping our behaviour. By doing so, we can achieve greater success and fulfilment in our lives.

When environments are not conducive to learning, how efficient is the learning overall?

Chapter 1: Introduction to ABA and behavioural Principles

■ **Overview of ABA and its historical development**

Applied behaviour Analysis (ABA) is a field of study that uses the principles of behaviourism to understand and modify human behaviour. ABA has a long and fascinating history that stretches back over a century. The roots of ABA can be traced back to the work of Ivan Pavlov, a Russian psychologist who is famous for his experiments with dogs and classical conditioning. Pavlov's work helped to establish the basic principles of behaviourism, which emphasise the importance of environmental stimuli in shaping behaviour. In the early 20th century, American psychologist John Watson expanded on Pavlov's work and helped to establish behaviourism as a distinct field of study. Watson believed that psychology should focus exclusively on observable behaviour, rather than inner mental processes such as thoughts and emotions. However, it was the work of American psychologist B.F. Skinner that truly revolutionised the field of behaviourism and laid the foundation for modern-day ABA. Skinner's experiments with operant conditioning demonstrated the power of reinforcement and punishment in shaping behaviour. Skinner's work led to the development of the first behaviour modification programs, which were used to treat a wide range of conditions such as phobias, addictions, and developmental disabilities. By the mid-20th century, behaviourism had gained mainstream acceptance and was being used in a variety of settings, from schools and hospitals to prisons and businesses.

In the 1960s and 1970s, the field of ABA began to emerge as a distinct branch of behaviourism. ABA focuses specifically on the use of behavioural principles to understand and modify human behaviour, with a particular emphasis on individuals with autism spectrum disorders. Today, ABA is a widely recognized and respected field, with applications in a wide and diverse range of people and settings. From improving academic performance, social skills, and reducing problem behaviours. These interventions can be applied in various settings, such as schools, homes, and community settings, to enhance the quality of life for individuals and their families. ABA has been shown to be an effective tool for shaping behaviour and improving outcomes. In the following sections of this chapter, we will delve deeper into the principles of ABA and how they are applied in practice. We will explore the role of reinforcement and punishment in behaviour modification, as well as the importance of functional behaviour assessments in identifying the underlying causes of problem behaviours. So join us as we continue our journey through the fascinating world of behaviourism and ABA.

Here are some of the classic Behavioural studies by Ivan Pavlov, John Watson, and B.F. Skinner and their relevance to everyday life:

Ivan Pavlov's research on classical conditioning is one of the most well-known and influential studies in the field of psychology. Pavlov discovered that dogs could be trained to associate a neutral stimulus, such as a bell, with a response that was previously only elicited by an unconditioned stimulus, such as food. This process, known as classical conditioning, demonstrated the power of environmental stimuli in shaping behaviour. Pavlov's findings have been applied in a variety of settings, including dog training and advertising. For example, dog trainers use classical conditioning to teach dogs to associate a command, such as "sit," with a reward, such as a treat. Similarly, advertisers use classical conditioning to create positive associations between a product and a particular stimulus, such as a catchy jingle or a beautiful model.

Behaviourism believes that all behaviours from all species can be described, predicted and controlled under relevant conditions

John Watson is considered the founder of behaviourism, a school of psychology that emphasises the study of observable behaviour and downplays the importance of mental processes such as thoughts and emotions. Watson's famous "Little Albert" experiment demonstrated the power of classical conditioning in shaping emotional responses. In this study, Watson conditioned a young child to fear a white rat by pairing the rat with a loud noise. The child eventually began to fear not only the rat, but other similar objects as well. This research has been applied in fields such as phobia treatment and exposure therapy. For example, therapists use classical conditioning to help individuals overcome fears by gradually exposing them to the feared object or situation in a safe and controlled environment.

B.F. Skinner's experiments with operant conditioning have had a profound impact on the field of behaviourism. Skinner discovered that behaviour could be shaped through reinforcement and punishment, and developed the concept of the Skinner box, a controlled environment for studying animal behaviour. Skinner's research has been applied in a wide range of settings, including education, parenting, and business management. For example, educators use reinforcement techniques to motivate students to engage in desired behaviours, such as completing homework or participating in class discussions. Parents use reinforcement and punishment to shape their children's behaviour, such as by rewarding good grades or taking away privileges for misbehaviour. Business managers use reinforcement techniques to increase productivity and reduce absenteeism, such as by offering bonuses for meeting sales targets or providing time off for good attendance.

All in all , the classic studies by Pavlov, Watson, and Skinner have had a profound impact on our understanding of human behaviour and how it can be modified. They have helped us to recognize the power of environmental stimuli in shaping our actions, and have provided practical tools for changing behaviour in a variety of contexts. From dog training to phobia treatment to workplace productivity, the principles discovered in these studies have relevance to our everyday lives. By applying these principles in our own lives and in the lives of those around us, we can make positive changes and improve our overall well-being.

■ Basic concepts of behaviourism, including operant and respondent conditioning, reinforcement, punishment, and extinction

behaviourism is a field of study that focuses on the role of environmental stimuli in shaping human behaviour. At its core, behaviourism is based on the idea that all behaviour is learned and can be modified through the use of reinforcement and punishment. There are several basic concepts that are central to behaviourism, including operant and respondent conditioning, reinforcement, punishment, and extinction. In this chapter, we will explore each of these concepts in detail and provide real-life examples of how they are applied in practice.

Reinforcement

Reinforcement is any consequence that increases the likelihood of a behaviour being repeated in the future. Positive reinforcement involves adding a desirable stimulus, such as a reward, in response to a behaviour. Negative reinforcement involves removing an aversive stimulus, such as taking away a painful electric shock, in response to a behaviour. A real-life example of positive reinforcement is a parent who rewards a child with a toy or treat for completing their homework. A real-life example of negative reinforcement is a person who takes an aspirin to relieve a headache and finds that the headache goes away. In a study conducted by psychologist Edward Deci, college students who were given a monetary reward for completing a puzzle were less likely to continue working on the puzzle once the reward was removed. This example illustrates the importance of intrinsic motivation in shaping behaviour. In a classic study, Skinner trained rats to press a lever by rewarding them with food pellets. He found that the rate of lever-pressing increased when the rewards were delivered on a variable rather than a fixed schedule. This example demonstrates the power of intermittent reinforcement in shaping behaviour.

Getting praise or an incentive that we really want can increase the future probability of doing that same behaviour under the same conditions.

Punishment

Punishment is any consequence that decreases the likelihood of a behaviour being repeated in the future. Positive punishment involves adding an aversive stimulus, such as a spanking, in response to a behaviour. Negative punishment involves removing a desirable stimulus, such as taking away a toy, in response to a behaviour. A real-life example of positive punishment is a teacher who reprimands a student for talking in class. A real-life example of negative punishment is a parent who takes away a child's phone for breaking a rule. A study conducted by psychologist Albert Bandura found that children who watched a film of an adult model hitting a Bobo doll were more likely to hit the doll themselves. This example illustrates the effects of observational learning and punishment. In a study conducted by psychologists Karen Pryor and Rosales Ruiz, students who were given a shock when they made a mistake in a learning task were more likely to make errors and less motivated to learn. This example demonstrates the negative effects of punishment in shaping behaviour.

Extinction

Extinction occurs when a behaviour that was previously reinforced is no longer reinforced and therefore decreases in frequency. This process is often used to eliminate problem behaviours. A real-life example of extinction is a parent who stops responding to a child's tantrums. If the tantrums are no longer reinforced by attention or other rewards, they will eventually stop occurring. In a study conducted by psychologist Alan Baron, the rate of gambling on slot machines decreased when the machines were modified to no longer dispense coins as rewards. This example demonstrates the effects of extinction in shaping behaviour. In a famous case study, Pavlov was able to extinguish a conditioned fear response in a child by gradually exposing him to the feared stimulus. This example illustrates the effectiveness of exposure therapy in shaping behaviour.

Operant Conditioning:

Operant conditioning is a type of learning that occurs when a behaviour is followed by a consequence. If the consequence is positive, such as a reward or reinforcement, the behaviour is more likely to be repeated in the future. If the consequence is negative, such as a punishment, the behaviour is less likely to be repeated. A classic example of operant conditioning is the Skinner box, a controlled environment for studying animal behaviour. In the Skinner box, an animal is placed in a box with a lever that, when pressed, releases a food pellet. Through trial and error, the animal learns that pressing the lever results in a positive consequence and will repeat the behaviour in the future. One of the most famous studies on operant conditioning was conducted by B.F. Skinner, who used a Skinner box to investigate the behaviour of rats and pigeons. In Skinner's experiments, an animal was placed in the box with a lever that, when pressed, released a food pellet. Through trial and error, the animal learned that pressing the lever resulted in a positive consequence and was more likely to repeat the behaviour in the future. This process of shaping behaviour through reinforcement is a central concept in operant conditioning.

In the early 1900s, Skinner also investigated the effects of punishment on behaviour. In one experiment, he placed a rat in a box and delivered a mild electric shock when the rat pressed a lever. The rat eventually learned to avoid pressing the lever altogether, demonstrating the effectiveness of punishment in shaping behaviour. In a study conducted by psychologists David Premack and Guy Woodruff, children were more likely to eat their vegetables when they were promised a desirable activity, such as watching TV, as a reward. This example demonstrates the power of reinforcement in shaping behaviour.

Respondent Conditioning:

Respondent conditioning is a type of learning that occurs when a neutral stimulus (e.g.a benign sensation, object, situation or person) is paired with a stimulus that naturally elicits a response (e.g.a particular sensation, object, situation or person that can triggers a feeling, behaviour or thought). Over time, the neutral stimulus comes to elicit the same response as the natural stimulus (e.g. smelling vanilla candles when you're out and about reminds you of your aunty and her vanilla candles). The gist is, something in your environment really reminds you of something or someone else.

A real-life example of respondent conditioning is a person who develops a phobia of flying after experiencing turbulence during a flight. The turbulence is a stimulus that elicits fear, while sitting in a plane itself is the neutral event. Over time, the person may come to fear getting in a plane even when there is no turbulence present. In a study conducted by psychologist Martin Seligman, dogs were subjected to repeated shocks that they could not control. The dogs eventually stopped trying to escape the shocks, even when a way out was available. This phenomenon, known as learned helplessness, demonstrates the powerful effects of respondent conditioning.

■ **The role of the environment in shaping behaviour**
behaviourism is a field of study that emphasises the role of environmental factors in shaping human behaviour. Our environment includes all of the physical, social, and cultural factors that surround us and influence our behaviour. The principles of behaviourism suggest that all behaviour, both positive and negative, is learned through the interaction of the individual with their environment. This process of learning occurs through the reinforcement or punishment of behaviours, as well as through the shaping of behaviours through environmental cues. Environmental factors can have a profound impact on our behaviour. For example, research has shown that children who grow up in poverty are more likely to experience academic and social difficulties later in life. This is due in part to the negative impact that poverty can have on the child's environment, including limited access to educational resources and social support. Another example of the impact of the environment on behaviour is the phenomenon of social contagion. This occurs when behaviours or attitudes spread rapidly through a social network, often without the individuals involved being aware of it.

One famous example of social contagion is the outbreak of mass hysteria in a small town in Tanzania in the 1960s. The outbreak was triggered by a single individual who claimed to have seen a ghost, and it quickly spread throughout the town, causing hundreds of people to experience similar hallucinations.

The principles of operant and respondent conditioning also demonstrate the role of the environment in shaping behaviour. For example, research has shown that individuals who are raised in environments that are rich in positive reinforcement, such as praise and encouragement, are more likely to exhibit positive behaviours later in life. Similarly, individuals who experience punishment or negative reinforcement in their environment are more likely to exhibit negative behaviours. In addition to the role of the environment in shaping behaviour, behaviourism also emphasises the importance of the environment in maintaining behaviour. This is known as the concept of discriminative control, which refers to the environmental cues that signal when a particular behaviour is appropriate or not. For example, a person who only smokes when they are around certain friends or in certain locations is exhibiting behaviour that is under discriminative control.

Behaviourism emphasises the importance of the environment in shaping verbal behaviour. Verbal behaviour refers to the use of language to communicate, and behaviourism suggests that verbal behaviour is learned through the same processes of reinforcement and punishment as nonverbal behaviour. For example, a child who is praised for using correct grammar is more likely to continue to use correct grammar in the future. To conclude, the principles of behaviourism suggest that the environment plays a crucial role in shaping and maintaining behaviour. By understanding the ways in which our environment influences our behaviour, we can take steps to modify our environment in positive ways and improve our lives.

Seeing others get benefits that we might want, could encourage us to emulate what others are doing.

As we learned, behaviourism emphasises the role of environmental factors in shaping human behaviour. Our environment includes all of the physical, social, and cultural factors that surround us and influence our behaviour. One important way in which the environment shapes behaviour is through culture. Culture is a set of shared values, beliefs, norms, and practices that shape the behaviour of individuals within a society. Different cultures have different norms and expectations for behaviour, which are learned through socialisation. For example, in some cultures, it is customary to bow or nod to show respect, while in others, a handshake or hug may be more appropriate. These cultural norms are learned through exposure to the environment and the social cues of others. Another example of how the environment shapes behaviour is through language acquisition. Children learn language through exposure to spoken

21

language in their environment, and through reinforcement and punishment of their verbal behaviour. For example, a child who is praised for correctly pronouncing a word is more likely to repeat that behaviour in the future.

The structure of the language itself, such as the grammar and syntax, can also shape behaviour. For example, languages that have a more formal structure may influence individuals to exhibit more formal behaviour in social situations. Parenting is another important factor in shaping behaviour. Parents who are warm and supportive, and who provide positive reinforcement for good behaviour, are more likely to raise children who exhibit positive behaviours. Conversely, parents who are harsh and punitive, and who provide negative reinforcement or punishment for bad behaviour, may raise children who exhibit negative behaviours. This can include behaviours such as aggression, substance abuse, and other problem behaviours. The education system is also designed to shape behaviour through the reinforcement of certain behaviours and the punishment of others. For example, students who do well on tests may be rewarded with good grades, while students who misbehave may be punished with detention or other consequences. The structure of the classroom environment, such as the seating arrangement or the use of technology, can also influence behaviour. For example, a classroom that is designed to facilitate collaboration and interaction may influence students to exhibit more social and cooperative behaviour. Media is another important factor in shaping behaviour.

The media we consume can have a significant impact on our behaviour, as it provides cues for behaviour through the portrayal of certain behaviours as desirable or undesirable. For example, a television show that depicts smoking as glamorous and cool may influence young people to take up the behaviour. Similarly, social media platforms that emphasise appearance and popularity may influence individuals to engage in certain behaviours to gain approval from others. In addition to these examples, the principles of operant and respondent conditioning also illustrate the role of the environment in shaping behaviour. Environmental factors can have a profound impact on our behaviour, as the reinforcement or punishment of behaviours, as well as the shaping of behaviours through environmental cues, can have a significant impact on our behaviour.

The environment plays a crucial role in shaping and maintaining behaviour. By understanding the ways in which our environment influences our behaviour, we can take steps to modify our environment in positive ways and improve our lives. This includes changing the way we parent, designing more effective educational systems, creating media that promotes positive behaviours, and more. By taking a behaviouristic approach to understanding the role of the environment in shaping behaviour, we can create a better world for ourselves and for future generations.

♆● The earliest recorded example of behavioural modification can be traced back to ancient Greece, where the philosopher Aristotle described a method for training horses based on the principles of reinforcement. Aristotle believed that the key to training a horse was to provide it with positive reinforcement for desirable behaviours and to use punishment for undesirable behaviours. He also believed that horses could be trained to respond to verbal cues and that they could be taught to perform complex tasks through a process of shaping behaviour. Aristotle's method of horse training was based on the principles of operant conditioning, which is the process of learning through the reinforcement of behaviours. Positive reinforcement involves providing a reward for a desirable behaviour, while negative reinforcement involves removing an aversive stimulus to encourage a desirable behaviour.

Punishment involves providing an aversive stimulus to discourage undesirable behaviour. The principles of operant conditioning have been applied to a wide range of human and animal behaviours, including addiction, phobias, and criminal behaviour. Aristotle's method of horse training was also based on the process of shaping behaviour, which involves breaking down a complex behaviour into smaller, more manageable steps and reinforcing each step until the behaviour is fully developed. Shaping behaviour is often used in animal training, such as teaching a dog to roll over or a dolphin to perform a backflip. It is also used in human behaviour modification, such as teaching a child to tie their shoes or an adult to quit smoking.

The principles of reinforcement and shaping have been applied to a wide range of behaviours beyond horse training. For example, they have been used to treat anxiety disorders, such as phobias, through a process known as exposure therapy. Exposure therapy involves exposing the individual to the feared stimulus in a controlled environment and providing positive reinforcement for calm behaviour. Similarly, the principles of reinforcement and shaping have been used in addiction treatment, such as providing rewards for remaining sober and gradually reducing the amount of reinforcement over time. Aristotle's method of horse training is a testament to the enduring value of the principles of operant conditioning and shaping behaviour. These principles have been applied across a wide range of behaviours, from animal training to human behaviour modification. They offer a powerful tool for shaping and modifying behaviour, whether it is in the context of personal growth, education, or therapy. And while the specifics of Aristotle's method may have been lost to time, the principles he outlined continue to inform our understanding of behaviour and how it can be shaped and modified to improve our lives.

Even as stoic Aristotle is, he understood the importance of learning via reinforcement

■ Environment Is Everything

In radical behaviourism, the environment is seen as everything that affects an individual's behaviour, including internal sensations and thoughts. This means that not only external factors, but also internal factors such as emotions, thoughts, and physiological responses, can influence and shape behaviour.

For example, if an individual experiences anxiety in social situations, this internal sensation may influence their behaviour in that context. They may avoid social situations, exhibit nervous behaviours such as fidgeting or avoiding eye contact, or engage in safety behaviours such as rehearsing what to say beforehand. In this case, the internal sensation of anxiety is an environmental factor that shapes the individual's behaviour.

Similarly, thoughts can also be seen as environmental factors that influence behaviour. For instance, if an individual has negative thoughts about their ability to succeed in a certain task, this may influence their behaviour in that context. They may avoid the task, procrastinate, or engage in self-defeating behaviours. In this case, the negative thoughts are an environmental factor that shapes the individual's behaviour.

Everything we experience, even a simple thought is within our environment

Chapter 2: The ABCs of behaviour

■ **Explanation of the ABCs of behaviour (Antecedent, behaviour, Consequence)**

The ABCs of behaviour, which stand for Antecedent, behaviour, and Consequence, are a set of principles that are central to the science of behaviourism. They help us understand the relationship between environmental events and behaviour, and how behaviour is shaped by the consequences that follow it. Antecedent refers to the events or stimuli that occur immediately before a behaviour. Antecedents can include anything in our environment that triggers a behaviour, such as a loud noise, a particular smell, an internal sensation or thought, or a specific action by another person. Antecedents can also include internal stimuli, such as thoughts, feelings, and physiological states. For example, a person might feel anxious before giving a speech, which can trigger nervous behaviours like fidgeting or avoiding eye contact. behaviour refers to the observable and measurable actions or responses of an individual. behaviours can include anything from physical actions like walking or talking to emotional responses like crying or laughing. In the context of behaviourism, behaviours are seen as a product of environmental factors, such as antecedents and consequences. Consequence refers to the events or stimuli that occur immediately after a behaviour.

Consequences can be either reinforcing or punishing. Reinforcing consequences increase the likelihood that a behaviour will be repeated in the future, while punishing consequences decrease the likelihood that a behaviour will be repeated. Reinforcing consequences can be positive, such as receiving a reward, or negative, such as the removal of an aversive stimulus. Punishing consequences can be positive, such as receiving a punishment, or negative, such as the removal of a positive stimulus.

Understanding the ABCs of behaviour can be useful in a variety of contexts, including education, parenting, and personal growth. For example, in education, teachers can use the ABCs to help students learn new behaviours by providing reinforcement for desirable behaviours and punishment for undesirable behaviours. In parenting, parents can use the ABCs to shape their child's behaviour by providing positive reinforcement for good behaviour and negative reinforcement or punishment for bad behaviour. In personal growth, individuals can use the ABCs to identify and change negative behaviours by understanding the environmental factors that trigger them and the consequences that follow. In the context of modern-day shopping, the ABCs of behaviour can be seen in the way that retailers use environmental factors to shape consumer behaviour. Antecedents can include anything in the shopping environment that triggers a behaviour, such as the layout of the store, the placement of products, and the use of sales and promotions. behaviours can include anything from browsing and comparing prices to making a purchase. Consequences can include rewards like discounts or loyalty points for making a purchase, or punishment like long lines or out-of-stock items for not making a purchase. By understanding the ABCs of behaviour, consumers can become more aware of how their behaviour is being shaped by the environment and the consequences that follow. This can help them make more informed decisions about their purchases and avoid behaviours that may be harmful or counterproductive. It can also help them identify and change negative behaviours by understanding the environmental factors that trigger them and the consequences that follow.

Ultimately, by understanding the ABCs of behaviour, consumers can become more empowered and in control of their shopping behaviour. The ABCs of behaviour (Antecedent, behaviour, and Consequence) can be used to analyse behaviour in a variety of everyday situations. Here are some more examples:

1. Social anxiety: Antecedent – You are invited to a party; behaviour – You feel nervous and avoid going or leave early; Consequence – You miss out on social connections and experiences. By analysing the ABCs of this behaviour, you can understand the antecedents that trigger your social anxiety, the behaviours that result, and the consequences that follow.

2. Insomnia: Antecedent – You have trouble falling asleep; behaviour – You start watching TV or scrolling through social media on your phone; Consequence – You stay up later than intended and feel tired the next day. By analysing the ABCs of this behaviour, you can understand the antecedents that lead to your insomnia, the behaviour itself, and the consequences that follow.

3. Nail biting: Antecedent – You feel anxious or bored; behaviour – You start biting your nails; Consequence – You feel a temporary sense of relief from the anxiety, but may experience negative consequences like nail damage or infection. By analysing the ABCs of this behaviour, you can understand the antecedents that trigger your nail biting, the behaviour itself, and the consequences that follow.

Cant stop wont stop. Nail biting is a common pastime for many, but have you thought about why? Sensation? Relaxation? Procrastination?

4. Smoking addiction: Antecedent – You feel stressed or anxious; behaviour – You smoke a cigarette; Consequence – You experience temporary relief from the stress, but may experience negative consequences like health problems or social isolation. By analysing the ABCs of this behaviour, you can understand the antecedents that trigger your smoking behaviour, the behaviour itself, and the consequences that follow.

5. Exercise avoidance: Antecedent – You feel tired or lazy; behaviour – You put off exercising; Consequence – You experience negative consequences like decreased fitness or weight gain. By analysing the ABCs of this behaviour, you can understand the antecedents that lead to your exercise avoidance, the behaviour itself, and the consequences that follow.

6. Productivity slump: Antecedent – You have a long to-do list; behaviour – You procrastinate by checking social media or watching TV; Consequence – You feel overwhelmed and behind schedule. By analysing the ABCs of this behaviour, you can understand the antecedents that lead to your productivity slump, the behaviour itself, and the consequences that follow.

7. Phobia response: Antecedent – You encounter a feared stimulus like a spider or flying; behaviour – You experience a physical response like sweating, shaking, or panic; Consequence – You avoid the feared stimulus or feel embarrassed about the physical response. By analysing the ABCs of this behaviour, you can understand the antecedents that trigger your phobia response, the behaviour itself, and the consequences that follow.

8. Mindless eating: Antecedent – You are distracted by the TV or your phone; behaviour – You eat without paying attention to your hunger cues or portion sizes; Consequence – You may overeat and experience negative health consequences. By analysing the ABCs of this behaviour, you can understand the antecedents that lead to mindless eating, the behaviour itself, and the consequences that follow.

These examples illustrate how the ABCs of behaviour can be used to understand a wide range of behaviours in everyday life. By analysing the antecedents, behaviours, and consequences of our own behaviours, we can gain a greater understanding of why we behave the way we do and how we can modify our behaviour to achieve our goals and improve our lives.

■ **Finding the Function (of Behaviour)**
In simple terms, every behaviour we engage in serves a purpose or has a reason behind it. This purpose or reason is known as the function of the behaviour. Think of it like this: just as every tool in a toolbox has a specific function to help us with a task, every behaviour we exhibit has a function to help us achieve something or respond to something in our environment.

For example, let's say someone frequently interrupts others during conversations. The function of this behaviour might be to gain attention or assert dominance in the conversation. On the other hand, if someone constantly avoids social situations, the function of this behaviour might be to reduce anxiety or escape from uncomfortable social interactions.

By understanding the function of a behaviour, we can gain insights into why it occurs and what purpose it serves for the individual. This understanding allows us to develop strategies and interventions that address the underlying needs or motivations driving the behaviour, ultimately leading to more effective behaviour change.

So, in a nutshell, every behaviour has a function, just like every tool has a purpose. By identifying and understanding the function of a behaviour, we can better understand why it happens and work towards modifying it in a positive and meaningful way.

■ The importance of understanding the ABCs in behaviour modification

Understanding the ABCs (Antecedent, behaviour, Consequence) of behaviour modification is critical for achieving meaningful and lasting behavioural change. By analysing the antecedents that trigger specific behaviours, the behaviours themselves, and the consequences that follow, we can gain a deeper understanding of why we behave the way we do and how we can modify our behaviour to achieve our goals. The importance of understanding the ABCs of behaviour modification lies in the fact that it helps us to identify and modify the antecedents and consequences that are responsible for problematic behaviours. By focusing on modifying these specific environmental factors, rather than relying solely on willpower or self-control, we can achieve more effective and lasting behaviour change. One way that the ABCs can be used to modify behaviour is by identifying and modifying antecedents that trigger problematic behaviours. For example, if you tend to eat unhealthy foods when you're stressed, identifying these triggers and finding alternative coping strategies can help you avoid the behaviour and achieve better health outcomes.

Another way that the ABCs can be used to modify behaviour is by identifying and modifying reinforcing consequences that maintain problematic behaviours. For example, if you tend to procrastinate by watching TV or checking social media, recognizing the temporary relief or pleasure that comes from these behaviours can help you find alternative sources of reinforcement that are more aligned with your goals. By modifying the antecedents and consequences of specific behaviours, we can create a more supportive and reinforcing environment for behaviour change. This approach is much more effective than simply relying on willpower or self-control, as it allows us to modify specific environmental factors that influence behaviour.

Additionally, understanding the ABCs of behaviour modification can help us to better understand our own behaviours and the behaviours of others. By analysing the ABCs of behaviour, we can gain a deeper understanding of why people behave the way they do, and how we can modify our own behaviours to achieve more meaningful and lasting change.

■ Real-world examples of the ABCs in action

1. Antecedent: Feeling tired or sleepy. behaviour: Go to bed. Consequence: You fall asleep and wake up rested.

2. Antecedent: Feeling stressed or anxious. behaviour: Going for a run. Consequence: Temporary relief from stress or anxiety.

3. Antecedent: Feeling bored or unstimulated. behaviour: Playing a video game. Consequence: Temporary sense of excitement or stimulation.

4. Antecedent: Feeling hungry. behaviour: Cooking a healthy meal. Consequence: Feeling satisfied and energised after eating the nutritious meal.

5. Antecedent: Feeling overwhelmed by a task. behaviour: Breaking the task down into smaller, more manageable steps. Consequence: Feeling less stressed and more confident about completing the task.

6. Antecedent: Feeling angry or frustrated. behaviour: Taking deep breaths and practising relaxation techniques. Consequence: Feeling more calm and in control of emotions.

7. Antecedent: Feeling lonely or isolated. behaviour: Attending a social gathering or event. Consequence: Feeling a sense of social connection and belonging.

Did you know out of the 233,360 hours we spend asleep, we tend to keep the environments the same. Bed, warmth, shelter, security etc.

8. Antecedent: Feeling stressed about finances. behaviour: Creating a budget and sticking to it. Consequence: Feeling more in control of finances and less stressed.

9. Antecedent: Feeling overwhelmed by a messy home. behaviour: Tidying up and organising living spaces. Consequence: Feeling less stressed and more in control of the environment.

10. Antecedent: Feeling anxious about a job interview. behaviour: Practising interview skills and rehearsing answers. Consequence: Feeling more confident and prepared for the interview.

These examples illustrate how the ABCs or the term contingency are present in various aspects of everyday life and can help us achieve more positive outcomes. By analysing the antecedents, behaviours, and consequences of specific behaviours, we can identify problematic behaviours and find alternative strategies that lead to more positive outcomes.

Chapter 3:How Motivation Operates

■ **Definition and explanation of motivating operations and how they influence behaviour**
Motivating operations (MOs) are environmental factors that temporarily alter the value of a particular consequence and thereby influence behaviour. MOs can be defined as events or conditions that increase or decrease the effectiveness of a particular consequence as a reinforcer or punisher. The concept of MOs is critical to understanding behaviour, as they influence not only the frequency of behaviour, but also the strength and direction of behaviour. One type of MO is an establishing operation (EO), which increases the value of a particular consequence as a reinforcer. An EO is an environmental event or condition that temporarily increases the value of a particular consequence and thereby increases the frequency or strength of behaviour that has been reinforced by that consequence. For example, if a person is hungry, food becomes more valuable as a reinforcer, and the person may be more likely to engage in behaviours that have been reinforced by food in the past. Another type of MO is an abolishing operation (AO), which decreases the value of a particular consequence as a reinforcer. An AO is an environmental event or condition that temporarily decreases the value of a particular consequence and thereby decreases the frequency or strength of behaviour that has been reinforced by that consequence. For example, if a person is no longer thirsty, water becomes less valuable as a reinforcer, and the person may be less likely to engage in behaviours that have been reinforced by water in the past.

MOs can also influence the effectiveness of punishment. A condition that makes a punishment more effective is known as an establishing operation for punishment (EOP), while a condition that makes a punishment less effective is known as an abolishing operation for punishment (AOP). MOs are critical to understanding behaviour, as they can influence the effectiveness of reinforcement and punishment. They can also explain why the same consequence can be reinforcing or punishing at different times, depending on the individual's current state or situation. For example, a particular food may be reinforcing when a person is hungry, but punishing when the person is already full.

Several prominent studies have explored the concept of MOs. In one classic study, DeLeon and Iwata (1996) examined the effects of an EO for attention on self-injurious behaviour in individuals with developmental disabilities. The study found that when the participants were given constant attention self-injurious behaviour (SIB) decreased. However, when attention was not available, the behaviour Increased. The findings suggest that when the individual had been deprived of attention (aka EO for attention), the individual engaged in behaviours that draw attention I.e SIB. This study demonstrated the importance of understanding the MOs that influence behaviour, as they can significantly impact the effectiveness of reinforcement and punishment.

Another study by Peterson et al. (2013) explored the effects of an EO for social interaction on the behaviour of individuals with autism. The study found that when the participants were given access to social interaction, their problem behaviours decreased. However, when social interaction was not available, the problem behaviours increased. The results indicate that when the individual was deprived of social interaction (aka EO for social interaction), the individual engaged in behaviours that may gain attention and thus, social interaction. This study highlights the importance of understanding MOs in designing effective behaviour interventions for individuals with autism.

All in all, motivating operations are critical to understanding behaviour, as they influence the effectiveness of reinforcement and punishment. Understanding the concept of MOs can help us design effective behaviour interventions and make sense of why behaviour can be reinforcing or punishing at different times. The next chapter will focus on the concept of discriminative control of behaviour and its role in shaping behaviour.

The marathon for success is an EO for success. Such a EO could have been triggered by our past experiences (e.g. growing up in a poor environment)

■ **Examples of how different types of motivating operations, such as establishing operations and abolishing operations, can affect behaviour**

1. Establishing operations for money: Feeling financially insecure or having a high value for money can serve as an EO for money, increasing the value of money as a reinforcer. This can lead to an increase in behaviours that have been reinforced by money in the past, such as working overtime or pursuing a high-paying job.

2. Abolishing operations for technology: Feeling overwhelmed by technology or having access to too much technology can serve as an AO for technology, decreasing the value of technology

as a reinforcer. This can lead to a decrease in behaviours that have been reinforced by technology in the past, such as scrolling through social media or playing video games.

3. Establishing operations for praise from specific individuals: Receiving praise from specific individuals, such as a boss or a romantic partner, can serve as an EO for praise from those individuals, increasing the value of praise from them as a reinforcer. This can lead to an increase in behaviours that have been reinforced by praise from those individuals in the past, such as completing tasks or showing affection.

4. Abolishing operations for certain foods: Feeling full or having access to unappealing food options can serve as an AO for certain foods, decreasing the value of those foods as reinforcers. This can lead to a decrease in behaviours that have been reinforced by those foods in the past, such as avoiding junk food or choosing healthier options.

5. Establishing operations for sensory input: Feeling sensory deprived or having access to a preferred sensory input, such as a cosy blanket or a favourite scent, can serve as an EO for that sensory input, increasing the value of that input as a reinforcer. This can lead to an increase in behaviours that have been reinforced by that sensory input in the past, such as seeking out cosy environments or using aromatherapy.

6. Abolishing operations for physical touch: Feeling overstimulated can serve as an AO for physical touch, decreasing the value of physical touch as a reinforcer. This can lead to a decrease in behaviours that have been reinforced by physical touch in the past, such as hugs or other forms of affection.

7. Establishing operations for social status: Feeling the need for social status or having access to high-status symbols, such as designer clothing or luxury cars, can serve as an EO for social status, increasing the value of social status as a reinforcer. This can lead to an increase in behaviours that have been reinforced by social status in the past, such as seeking high-paying jobs or pursuing prestigious memberships.

8. Abolishing operations for risk-taking behaviour: Feeling anxious or having access to safer options can serve as an AO for risk-taking behaviour, decreasing the value of risk-taking behaviour as a reinforcer. This can lead to a decrease in behaviours that have been reinforced by risk-taking in the past, such as avoiding extreme sports or taking fewer risks in business.

9. Establishing operations for physical activity with others: Feeling the need for social interaction or having access to a group that engages in physical activity, such as a sports team or workout buddy, can serve as an EO for physical activity with others, increasing the value of physical activity with others as a reinforcer. This can lead to an increase in behaviours that have been reinforced by physical activity with others in the past, such as joining a sports team or working out with a friend.

10. Abolishing operations for material possessions: Feeling overwhelmed by material possessions or having access to fewer material possessions can serve as an AO for material possessions, decreasing the value of material possessions as reinforcers. This can lead to a decrease in behaviours that have been reinforced by material possessions in the past,

11. Establishing operations for intellectual stimulation: Feeling intellectually curious or having access to a challenging intellectual task can serve as an EO for intellectual stimulation, increasing the value of intellectual stimulation as a reinforcer. This can lead to an increase in behaviours that have been reinforced by intellectual stimulation in the past, such as reading a challenging book or engaging in a stimulating conversation.

12. Abolishing operations for sleepiness: Feeling awake and alert or having access to caffeine can serve as an AO for sleepiness, decreasing the value of sleepiness as a punisher. This can lead to a decrease in behaviours that have been punished by sleepiness in the past, such as falling asleep during a lecture or feeling lethargic during work.

13. Establishing operations for praise from a certain group: Receiving praise from a certain group, such as a peer group or a specific social circle, can serve as an EO for praise from that group, increasing the value of praise from them as a reinforcer. This can lead to an increase in behaviours that have been reinforced by praise from that group in the past, such as seeking approval or validation from them.

14. Abolishing operations for fear: Feeling safe and secure or having access to a supportive environment can serve as an AO for fear, decreasing the value of fear as a punisher. This can lead to a decrease in behaviours that have been punished by fear in the past, such as avoiding certain situations or activities due to fear.

15. Establishing operations for access to technology: Feeling a need for technological access or having access to desirable technology can serve as an EO for access to technology, increasing the value of technology as a reinforcer. This can lead to an increase in behaviours that have been reinforced by technology in the past, such as checking social media or using electronic devices.

Sometimes having too much of something can usually be a turn off

16. Abolishing operations for social rejection: Feeling socially accepted or having access to supportive social relationships can serve as an AO for social rejection, decreasing the value of social rejection as a punisher. This can lead to a decrease in behaviours that have been punished by social rejection in the past, such as avoiding social situations or experiencing anxiety in social contexts.

17. Establishing operations for financial security: Feeling financially insecure or having access to financial resources can serve as an EO for financial security, increasing the value of financial security as a reinforcer. This can lead to an increase in behaviours that have been reinforced by financial security in the past, such as saving money or pursuing financial stability.

18. Abolishing operations for physical discomfort: Feeling physically comfortable or having access to physical comfort can serve as an AO for physical discomfort, decreasing the value of

physical discomfort as a punisher. This can lead to a decrease in behaviours that have been punished by physical discomfort in the past, such as avoiding uncomfortable physical sensations or activities.

19. Establishing operations for novelty: Feeling bored or having access to novel experiences can serve as an EO for novelty, increasing the value of novelty as a reinforcer. This can lead to an increase in behaviours that have been reinforced by novelty in the past, such as seeking out new experiences or engaging in novel activities.

20. Abolishing operations for work: Feeling overworked or having access to fewer work responsibilities can serve as an AO for work, decreasing the value of work as a reinforcer. This can lead to a decrease in behaviours that have been reinforced by work in the past, such as taking a break or reducing work hours.

Motivating operations (MOs) are factors that affect the value of a consequence as a reinforcer or punisher. They work in conjunction with the antecedent-behaviour-consequence (ABC) model of behaviour to explain how environmental variables can influence behaviour.

For example, an establishing operation (EO) can increase the value of a consequence as a reinforcer, making the behaviour more likely to occur. If someone is hungry (EO), the value of food as a reinforcer is increased, making the behaviour of eating more likely to occur. In this case, hunger serves as the EO, the behaviour is eating, and the consequence is the consumption of food. On the other hand, an abolishing operation (AO) can decrease the value of a consequence as a reinforcer, making the behaviour less likely to occur. If someone is already full (AO), the value of food as a reinforcer is decreased, making the behaviour of eating less likely to occur. In this case, the fullness serves as the AO, the behaviour is not eating, and the consequence is the absence of food consumption. The takeaway is that MOs help explain why the same consequence can function as a reinforcer or punisher depending on the person's current state or environment. By understanding the influence of MOs, behaviour analysts can better predict and modify behaviour.

■ Multiple MOs all the time

In the modern day, working adults often face multiple responsibilities and competing demands on their time and resources. This can create a complex interplay of motivating operations (MOs) that can affect behaviour in unpredictable ways. For example, a working parent may be motivated to earn a high salary (an EO for financial security) while also feeling physically fatigued due to a lack of sleep (an EO for rest). These two MOs may compete with each other, leading to a behaviour that is a compromise between the two (such as working long hours to earn a high salary, but sacrificing sleep in the process). On the other hand, multiple MOs can also interact synergistically, amplifying each other's effects. For instance, a working adult may be motivated to complete a task before a deadline (an EO for escape from aversive consequences) while also feeling socially isolated due to a lack of social interaction (an EO for social attention). These two MOs may reinforce each other, leading to a behaviour that is more extreme than either MO alone (such as staying up late to complete a task and neglecting social relationships).

Moreover, the effects of MOs can vary depending on their strength and the specific situation. For instance, a working adult who is usually motivated to avoid physical discomfort may be more willing to endure it if they are in a situation where avoiding it is impossible or undesirable (such as during a challenging work project or to obtain a highly valued promotion). In order to effectively modify behaviour, it is important for behaviour analysts to take into account the interaction between multiple MOs in the context of a working adult's daily life. By identifying and manipulating MOs, analysts can help working adults achieve a balance between their various responsibilities and demands. This requires a thorough understanding of the specific MOs that are present and how they interact with each other, as well as the unique circumstances of the individual's life.

■ The role of motivating operations in behaviour modification

Motivating operations (MOs) play a crucial role in behaviour modification, as they affect the value of a consequence as a reinforcer or punisher. Without taking MOs into account, behaviour analysts may miss key factors that contribute to a person's behaviour, leading to less effective interventions. One major advantage of understanding MOs is that they help behaviour analysts predict and control behaviour. By identifying and manipulating MOs, analysts can increase the likelihood of desired behaviour and decrease the likelihood of undesired behaviour. For example, if a child is more likely to engage in problem behaviour when they are tired, a behaviour analyst can modify the environment by providing more opportunities for rest and relaxation, thereby reducing the influence of the EO for fatigue. Moreover, MOs can help explain why the same consequence can function as a reinforcer or punisher depending on the person's current state or environment. For instance, food may be a reinforcer for someone who is hungry (EO), but it may function as a punisher for someone who is already full (AO). Understanding this dynamic can help behaviour analysts tailor interventions to individual needs, maximising the effectiveness of behaviour modification. Additionally, understanding MOs is crucial in preventing problem behaviour. Many problem behaviours arise from a person's attempts to meet their own needs, such as escaping aversive situations or obtaining desired items. By identifying the MOs that contribute to problem behaviour, behaviour analysts can develop proactive interventions that provide alternative ways to meet those needs, reducing the likelihood of problem behaviour. Just remember that the role of MOs in behaviour modification cannot be overstated. By taking into account the influence of MOs, behaviour analysts can better understand and modify behaviour, leading to more successful interventions and better quality of life for the individuals they serve.

Chapter 4: Behavioural Control

■ Explanation of discriminative stimuli and their role in behaviour

Discriminative stimuli (SDs) are a crucial component of behaviour analysis and play a key role in shaping behaviour. In simple terms, SDs are cues or signals that indicate the availability of a specific consequence following a behaviour. This can include anything from a sound or visual cue to a particular setting or context. For example, imagine that a child is given a sticker every time they clean up their toys. Over time, the child may learn to associate the sound of a parent's voice saying "clean up" with the availability of the sticker. In this case, the sound of the parent's voice serves as the SD, signalling to the child that if they clean up their toys, they will receive the sticker as a consequence. SDs can be used to increase or decrease the likelihood of a particular behaviour, depending on the type of consequence that follows. For instance, an SD that signals the availability of a desirable consequence (such as a reward) can increase the likelihood of a behaviour, while an SD that signals the availability of an aversive consequence (such as punishment) can decrease the likelihood of a behaviour. Moreover, the effects of SDs can be influenced by a variety of factors, including the person's history of reinforcement and punishment, the specific context or setting, and the current motivational state. For example, a person who has a history of being rewarded for a behaviour may be more likely to respond to an SD that signals the availability of a reward, while a person who has a history of being punished for a behaviour may be more likely to respond to an SD that signals the avoidance of punishment. In order to effectively modify behaviour using SDs, behaviour analysts must carefully select and manipulate the relevant cues or signals to ensure that the desired behaviour is reinforced or punished consistently. This requires a thorough understanding of the individual's history of reinforcement and punishment, as well as the specific context in which the behaviour occurs.

So we can surmise that the concept of discriminative stimuli is a critical tool for behaviour analysts in shaping behaviour. By understanding the role of SDs in behaviour and carefully selecting and manipulating relevant cues or signals, behaviour analysts can increase the likelihood of desired behaviour and decrease the likelihood of undesired behaviour.

When we see a cue for reinforcement that we know will work..we respond accordingly!

■ **Examples of how discriminative stimuli can influence behaviour**

Discriminative stimuli (SDs) are cues or signals that indicate the availability of a specific consequence following a behaviour. SDs play a key role in shaping behaviour, as they can increase the likelihood of a particular behaviour occurring in response to a specific stimulus. The following are examples of how SDs can influence behaviour in various settings:

1. Classroom behaviour:

In a classroom setting, a teacher may use specific cues or signals as SDs to indicate that students should stop talking, raise their hand, or pay attention. For example, a teacher may say "quiet" or ring a bell to signal the start of a lesson, or raise their hand to indicate that students should do the same. These cues serve as SDs, signalling to students that they should engage in a particular behaviour in response to the teacher's cue.

2. Sports Performance:

In sports, certain cues or signals can serve as SDs to help athletes perform specific actions more effectively. For example, a coach may use a whistle to signal the start or end of a play, or use a specific phrase to indicate that a particular play should be executed. These cues serve as SDs, signalling to the athletes that they should perform a particular behaviour in response to the coach's cue.

3. Medical Treatment:

In medical settings, certain cues or signals can serve as SDs to help patients comply with treatment plans. For example, a healthcare provider may use a specific sound or visual cue to signal that it is time for a patient to take their medication. These cues serve as SDs, signalling to the patient that they should engage in a particular behaviour in response to the healthcare provider's cue.

4. Household Chores:

In a home setting, certain cues or signals can serve as SDs to help family members complete household tasks. For example, a parent may use a specific sound or visual cue to signal that it is time for their child to clean their room. These cues serve as SDs, signalling to the child that they should engage in a particular behaviour in response to the parent's cue.

5. Addiction Recovery:

In addiction recovery, certain cues or signals can serve as SDs to help individuals avoid relapse. For example, a person in recovery may avoid certain people, places, or things that they associate with their past substance use. These cues serve as SDs, signalling to the individual that they should engage in a particular behaviour (avoidance) in response to the presence of certain cues or stimuli.

6. Food Choices:

In the context of food choices, certain cues or stimuli can serve as SDs that influence the selection of foods. For example, a person may be more likely to select a food item if it is packaged in a certain way or if it is displayed in a particular location. These cues serve as SDs, signalling to the person that they should engage in a particular behaviour (selecting a specific food item) in response to the presence of certain cues or stimuli.

7. Social Interactions:

In the context of social interactions, certain cues or stimuli can serve as SDs that influence behaviour. For example, a person may be more likely to initiate conversation or engage in a certain behaviour (e.g., smiling, making eye contact) in response to the presence of certain cues or stimuli (e.g., a friendly greeting, a positive facial expression). These cues serve as SDs, signalling to the person that they should engage in a particular behaviour in response to the presence of certain cues or stimuli.

8. Consumer behaviour:

In the context of consumer behaviour, certain cues or stimuli can serve as SDs that influence purchasing decisions. For example, a person may be more likely to purchase a product if it is packaged in a certain way, if it is on sale or if it is advertised in a particular way. These cues serve as SDs, signalling to the person that they should engage in a particular behaviour (e.g. purchasing the product) in response to the presence of said cues.

We all tend to behave more in line with the law if law enforcement are present

Discriminative stimuli can have a significant impact on individuals with behavioural agnosia. As these individuals are unable to recognize or interpret contextual cues, they may have difficulty discriminating between different stimuli and may engage in inappropriate behaviours.

Understanding the role of discriminative stimuli in behaviour can help in modifying and shaping behaviour in individuals with behavioural agnosia, and can improve their ability to function effectively in different environments. By recognizing and manipulating discriminative stimuli, behaviour analysts can help individuals with behavioural agnosia to acquire new skills and behaviours, and reduce problematic behaviours.

■ **The importance of understanding discriminative control in behaviour modification**
Discriminative control refers to the relationship between discriminative stimuli (SDs) and the behaviour that they evoke. Understanding this relationship is crucial for individuals who want to modify their behaviour in a desired way, as well as for professionals in fields such as education, healthcare, and marketing who seek to influence the behaviour of others. By understanding the role of discriminative stimuli in behaviour, individuals can gain greater control over their own behaviour. This can be particularly helpful when trying to break a bad habit or adopt a new positive behaviour. For example, if someone wants to quit smoking, they can identify the SDs that trigger their smoking behaviour (such as certain people or situations) and work to avoid or replace those stimuli. Similarly, if someone wants to start exercising regularly, they can identify the SDs that will prompt them to exercise (such as setting an alarm or laying out workout clothes the night before) and create a plan to ensure that those stimuli are present. For professionals in fields such as education, healthcare, and marketing, understanding discriminative control is essential for achieving desired outcomes. In educational settings, teachers can use SDs to prompt desired behaviours in their students, such as raising their hand before speaking or lining up quietly for recess. In healthcare settings, healthcare providers can use SDs to prompt patients to engage in health-promoting behaviours, such as taking their medication as prescribed or following a recommended diet. In marketing, companies can use SDs to influence consumer behaviour, such as by creating eye-catching packaging or by placing products in strategic locations in stores.

Furthermore, understanding discriminative control is crucial for ensuring the social validity of behaviour modification interventions. Social validity refers to the degree to which an intervention is perceived as acceptable, important, and relevant by those who are affected by it. By taking into account the role of discriminative stimuli in behaviour, behaviour modification interventions can be designed to be socially valid and respectful of the individual's values and preferences.

■ **Real-world examples of discriminative control in action**
Discriminative control is all around us, shaping our behaviour in a variety of settings. Here are several real-world examples of how discriminative control works in everyday life:

1. Driving: When you approach a red traffic light, it serves as a discriminative stimulus, signalling that you should stop your car. Conversely, a green traffic light serves as a discriminative stimulus, signalling that you should continue driving.

2. Socialising: When you attend a formal event, such as a wedding or job interview, the formal dress code serves as a discriminative stimulus, signalling that you should dress in a certain way.

In contrast, a casual social gathering, such as a backyard barbecue, might not have any specific dress code.

3. Eating: When you walk into a restaurant, the aroma of food serves as a discriminative stimulus, signalling that it is time to eat. The sight of a menu or the presence of a waiter might also serve as discriminative stimuli, signalling that it is time to order food.

4. Learning: In a classroom setting, the teacher's instruction and the presence of educational materials, such as textbooks or whiteboards, serve as discriminative stimuli, signalling that it is time to focus and learn.

5. Exercise: For people who want to establish a regular exercise routine, setting aside a specific time of day, such as early morning or after work, can serve as a discriminative stimulus, signalling that it is time to exercise.

6. Shopping: In a grocery store, items that are placed at eye-level and have bright, attractive packaging can serve as discriminative stimuli, prompting shoppers to purchase those items. Similarly, promotional displays and advertisements can serve as discriminative stimuli, prompting shoppers to make specific purchases.

7. Music: When you hear a particular song or type of music, it can serve as a discriminative stimulus, prompting you to engage in a particular behaviour, such as dancing or singing along.

8. Work: In a work setting, deadlines, meetings, and email notifications can serve as discriminative stimuli, prompting employees to complete specific tasks or shift their attention to a particular project.

In each of these examples, discriminative control plays a crucial role in shaping behaviour. By recognizing and utilising the power of discriminative stimuli, individuals and professionals can promote positive behaviour change and achieve desired outcomes.

We all know how deadlines can really push us, especially as time draws closer to it!

Research on discriminative stimuli has provided valuable insight into the ways that SDs influence behaviour. Here are several examples of illustrative research studies:

1. Waterman and Powell (1971): In this study, researchers trained pigeons to peck a key in response to a specific visual stimulus (a horizontal line). They then systematically varied the orientation of the line and observed the pigeons' response rates. The results showed that the response rate was highest when the line was presented at the trained orientation and decreased as the orientation deviated from the trained orientation. This study demonstrated the powerful influence of discriminative stimuli on behaviour.

2. Binder and Haughton (1967): In this study, researchers trained rats to press a lever in response to a specific auditory stimulus (a tone). They then presented the rats with a series of novel auditory stimuli and observed their response rates. The results showed that the rats were more likely to respond to stimuli that were similar in frequency to the trained stimulus than to stimuli that were dissimilar. This study demonstrated the importance of stimulus similarity in discriminative control.

3. Timberlake and Grant (1975): In this study, researchers trained rats to press a lever in response to a specific visual stimulus (a flashing light). They then presented the rats with a series of trials in which the light was presented either alone or in combination with a noise. The results showed that the presence of the noise disrupted the discriminative control of the light, indicating that the rats were attending to both stimuli. This study demonstrated the importance of stimulus control in discriminative behaviour.

4. Feinman and Glass (1976): In this study, researchers trained rats to press a lever in response to a specific visual stimulus (a vertical line). They then presented the rats with a series of trials in which the line was either presented alone or in combination with a visual distractor (a circle). The results showed that the presence of the distractor disrupted the discriminative control of the line, indicating that the rats were attending to both stimuli. This study demonstrated the importance of stimulus relevance in discriminative behaviour.

5. Guttman and Kalish (1956): In this classic study, researchers trained a child to select a specific object (a toy car) from a set of objects in response to a specific verbal stimulus ("Give me the car"). They then presented the child with a series of trials in which the verbal stimulus was presented with a different set of objects. The results showed that the child was able to generalise the discriminative stimulus to novel sets of objects, demonstrating the power of stimulus control in human behaviour.

Every principle we discussed so far is applicable to all animal life on earth!

Chapter 5: Verbal Behaviour

■ **Overview of verbal behaviour and its role in ABA**

Verbal behaviour is a subset of behaviour that involves the use of language and communication. It is a complex and multifaceted area of study that has significant implications for behaviour analysis and intervention. In the field of ABA, the study of verbal behaviour is crucial for understanding and shaping human behaviour. Verbal behaviour is characterised by three primary components: the speaker, the listener, and the context. The speaker is the individual who emits the verbal behaviour, while the listener is the individual who receives and interprets the verbal behaviour. The context refers to the environmental conditions that influence the verbal behaviour. One of the key concepts in verbal behaviour is the functional relationship between the verbal behaviour and its consequences. This relationship is described by the four-term contingency: the antecedent, the behaviour, the consequence, and the motivating operation. The antecedent is the stimulus or event that precedes the behaviour, while the consequence is the event that follows the behaviour. The motivating operation is the environmental condition that affects the value of a particular consequence and influences the likelihood of the behaviour occurring.

There are many different types of verbal behaviour, including manding, tacting, intraverbals, and autoclitics. Each type of verbal behaviour serves a unique function and has different requirements for acquisition and maintenance. For example, manding is a verbal request for a specific item or activity, while tacting is a verbal label for an object or event. Verbal behaviour is a crucial aspect of ABA, as it is used to teach and shape a wide range of skills and behaviours. For example, therapists may use manding to teach a child to request a desired item or activity, or may use tacting to teach a child to label objects or events in their environment. Verbal behaviour interventions have been shown to be effective in improving communication skills, reducing problem behaviours, and increasing social interactions. In conclusion, verbal behaviour is a critical component of behaviour analysis and ABA. It is a complex area of study that involves the use of language and communication and is characterised by the functional relationship between the verbal behaviour and its consequences. Understanding the principles of verbal behaviour is essential for behaviour analysts and therapists to effectively teach and shape behaviour.

The role of verbal behaviour in our daily lives cannot be overstated. Communication is an essential part of our existence, and verbal behaviour plays a crucial role in our ability to communicate effectively with others. From simple conversations with friends and family to more complex interactions in the workplace or academic settings, verbal behaviour is a necessary tool for social interaction and learning. In our personal lives, verbal behaviour is essential for expressing our wants, needs, and emotions. We use verbal behaviour to ask for help, to express our feelings, to make plans, and to negotiate with others.

Verbal behaviour is also critical for forming and maintaining relationships, as it allows us to connect with others and develop shared meanings and understandings In the workplace, verbal behaviour is essential for effective communication between colleagues and with customers. Clear and concise verbal behaviour can help to avoid misunderstandings and miscommunications, leading to increased productivity and job satisfaction. Verbal behaviour can also be used to motivate and encourage employees, as well as to provide constructive feedback and praise. In educational settings, verbal behaviour is a critical component of learning and academic success. Students must be able to understand and respond to verbal instructions, engage in classroom discussions, and demonstrate their understanding through verbal responses. Teachers also use verbal behaviour to provide feedback, give instructions, and facilitate learning.

■ **Explanation of the different types of verbal behaviour, such as manding, tacting, and intraverbals**

Verbal behaviour is a complex and multifaceted area of study that involves the use of language and communication. There are several different types of verbal behaviour, each of which serves a unique function and has different requirements for acquisition and maintenance. In this section, we will explore three of the primary types of verbal behaviour: manding, tacting, and intraverbals.

Manding

Manding is a type of verbal behaviour that involves making requests or demands for specific items, actions, or events. For example, a child might say "I want milk" to request a glass of milk, or "I need to go to the bathroom" to request a bathroom break. Manding is an important skill for individuals to acquire, as it allows them to communicate their wants and needs effectively.

Tacting

Tacting is a type of verbal behaviour that involves labelling or describing objects, events, or actions in the environment. For example, a child might say "that's a ball" when presented with a ball, or "it's raining outside" when looking out the window on a rainy day. Tacting is an essential skill for individuals to acquire, as it allows them to understand and interact with the world around them.

A child tacting a….???.. I don't even think the child knows..

Intraverbals

Intraverbals are a type of verbal behaviour that involves responding to verbal stimuli with a verbal response that is not necessarily related to the original stimulus. For example, a child might say "Happy birthday" in response to hearing the phrase "What do you say to your friend on their special day?" Intraverbals are a more complex type of verbal behaviour and require a higher level of cognitive and linguistic abilities.

Humans use these different types of verbal behaviour in a variety of contexts, from simple conversations with friends and family to more complex interactions in the workplace or academic settings. For example, in a social context, a person might use manding to request a drink at a party, tacting to describe the decor of a restaurant, or intraverbals to engage in a philosophical debate.

In educational settings, verbal behaviour is crucial for learning and academic success. Students use manding to request clarification on assignments, tacting to describe concepts in class, and intraverbals to engage in discussions and debates. The ability to use these types of verbal behaviour effectively can lead to improved relationships, increased learning opportunities, and greater overall success in personal and professional endeavours.

Certainly. Verbal behaviour is a complex topic with many different facets, but there are a number of studies and stories that can help illustrate the different types of verbal behaviour and their importance. One classic study on manding comes from research on children with autism. In this study, researchers taught a child to say "I want" followed by the name of a desired item. Over time, the child was able to use this skill to request a wide range of items and activities, improving their ability to communicate their needs effectively.

Tacting is another type of verbal behaviour that can be illustrated through stories and examples. For instance, a child learning to identify and label different colours might initially make many errors, but with practice and reinforcement, they may eventually be able to accurately identify and label colours in a variety of contexts.

Intraverbal behaviour can be illustrated through examples like the game "Name That Tune." In this game, one person hums a few bars of a song and another person has to guess the title or artist. This requires the listener to use their knowledge of music and verbal behaviour to come up with a response that is not necessarily related to the original stimulus.

Finally, there is a wide range of research on the use of augmentative and alternative communication (AAC) devices to teach children with autism and other language impairments to communicate effectively. One study found that children who received AAC interventions were more likely to engage in conversation and social interaction with others, leading to improvements in their overall quality of life.

These examples help to illustrate the different types of verbal behaviour and how they can be taught and used in everyday life. By understanding these principles and working to improve our own verbal behaviour skills, we can enhance our communication abilities and improve our interactions with others.

■ **The importance of understanding verbal behaviour in communication and behaviour modification**

Verbal behaviour plays a crucial role in communication and behaviour modification. Understanding the different types of verbal behaviour and how they function is essential for improving our communication skills and modifying behaviour in ourselves and others. Verbal behaviour refers to the use of language to communicate with others. It includes everything from speaking and writing to signing and gesturing. One of the key benefits of verbal behaviour is that it allows us to convey complex ideas and emotions, which would be difficult or impossible to communicate through nonverbal means alone. Verbal behaviour also plays a critical role in behaviour modification. By teaching individuals new ways to communicate their needs and desires, we can help them to become more independent and self-sufficient. For example, teaching a child with autism to use verbal behaviour to request a desired item can reduce frustration and improve their overall quality of life. Moreover, understanding verbal behaviour is important for behaviour modification because it allows us to identify and change the antecedents and consequences of behaviour.

By manipulating these environmental factors, we can influence how individuals behave and ultimately help them to achieve their goals. For example, by reinforcing appropriate verbal behaviour, we can encourage individuals to use these skills more frequently and in a wider range of situations. In addition to improving communication and behaviour modification, understanding verbal behaviour can also enhance our social interactions with others. Effective communication skills allow us to build stronger relationships, resolve conflicts more effectively, and better understand the needs and desires of others. Moreover, verbal behaviour can help us to convey our emotions and connect with others on a deeper level, leading to more fulfilling and satisfying social experiences.

By learning more about the different types of verbal behaviour and how they function, we can become more effective communicators and ultimately improve our overall quality of life.

■ **Real-world examples of verbal behaviour in action**
- A baby crying to communicate their need for food or attention
- An individual with autism learning to use a picture exchange communication system to request items
- A toddler learning to use words to express their emotions
- A teacher using positive reinforcement to encourage a student to participate in class discussions
- A salesperson using persuasive language to convince a customer to make a purchase
- A therapist using verbal prompts to teach a client new skills
- A politician using language to sway public opinion
- A comedian using language to elicit laughter from an audience
- A radio host engaging with listeners through verbal interaction
- A doctor using clear and concise language to explain a medical diagnosis to a patient.

Professional marketing and sales people are experts in using the right words to get you to do certain behaviours e.g. paying attention

Chapter 6: Verbal Communities

■ **Explanation of verbal communities and how they influence behaviour**

Verbal communities are groups of individuals who share a common language and set of verbal behaviours. These communities can range from families and social groups to entire cultures and societies. The language and verbal behaviour used within a community are shaped by a variety of factors, including history, culture, and social norms. The influence of verbal communities on verbal behaviour is significant. People within a particular community often develop a shared set of linguistic conventions, including vocabulary, grammar, and syntax. These conventions are reinforced and passed down through generations, ultimately shaping the way that individuals within that community use language. For example, individuals who grow up in a bilingual household may learn to switch back and forth between two languages, depending on who they are speaking with. Alternatively, individuals who grow up in a community where a specific dialect or accent is prominent may adopt those linguistic features in their own speech.

Verbal communities also play a significant role in shaping behaviour more broadly. The language and verbal behaviour used within a community can influence social norms and expectations, which in turn can shape individual behaviour. For example, if a community places a high value on academic achievement, individuals within that community may be more likely to prioritise education and pursue academic goals. Moreover, verbal communities can influence behaviour by providing opportunities for reinforcement and punishment. Individuals within a community may be more likely to receive social approval or disapproval for their behaviour, depending on how it aligns with the norms and values of the community. In addition to shaping behaviour, verbal communities can also influence attitudes and beliefs. For example, individuals who grow up within a particular religious community may adopt the beliefs and values of that community, influencing their behaviour and decision-making processes. By recognizing the influence of verbal communities on behaviour, we can better understand how language and social norms shape individual behaviour and work towards creating positive change within our communities.

■ **Examples of how verbal communities can shape behaviour and communication**
1. Language dialects and accents: People who grow up in a particular region or community often adopt the local dialect and accent in their speech. For example, someone who grows up in the Southern United States may adopt a Southern accent, while someone who grows up in New York City may adopt a New York accent. This can influence the way that individuals communicate and how they are perceived by others.

2. Social norms and expectations: Verbal communities can have a significant impact on social norms and expectations. For example, in some communities, it may be considered rude to interrupt others while they are speaking, while in others it may be perfectly acceptable. These norms and expectations can shape the way that individuals communicate with one another and can influence behaviour.

3. Religious beliefs and values: Verbal communities can also shape religious beliefs and values, which can influence behaviour. For example, someone who grows up within a particular religious community may adopt the values and beliefs of that community, shaping the way that they behave and the decisions that they make.

4. Education and academic achievement: Verbal communities can influence attitudes towards education and academic achievement. For example, in some communities, there may be a

strong emphasis on academic success and achievement, which can influence the behaviour of individuals within that community.

5. Group identity and belonging: Verbal communities can also shape group identity and the sense of belonging that individuals feel within a community. For example, individuals who grow up speaking a particular language may feel a sense of identity and belonging to that community, which can influence their behaviour and communication.

6. Gender roles and expectations: Verbal communities can also influence gender roles and expectations. For example, in some communities, there may be gender-specific language conventions or expectations regarding the way that men and women communicate.

7. Political beliefs and values: Verbal communities can also shape political beliefs and values, which can influence behaviour and communication. For example, individuals who grow up within a particular political community may adopt the values and beliefs of that community, influencing the way that they communicate about political issues.

Online, there are various different websites, social media, facebook groups, subreddits etc. that act as their own verbal communities.

■ **Examples of how verbal communities can shape behaviour and communication**

1. A study conducted by sociologist Elijah Anderson in the 1990s explored the use of language and nonverbal cues in a predominantly African American neighbourhood in Philadelphia. Anderson found that residents of the neighbourhood often used a specific type of language, referred to as "code-switching," in order to communicate effectively within their community. This code-switching involved the use of slang and nonstandard English, which allowed residents to convey complex social meanings and navigate complex social hierarchies within the community.

2. The concept of "linguistic relativity," popularly known as the Sapir-Whorf hypothesis, suggests that the language we speak can influence the way that we think and perceive the world around us. This hypothesis has been applied to a wide range of studies, including studies of cultural differences in communication and studies of the relationship between language and cognition.

3. A study conducted by psychologist Albert Bandura in the 1960s explored the role of verbal persuasion in shaping behaviour. Bandura found that individuals who were exposed to persuasive messages were more likely to adopt certain behaviours, even if those behaviours conflicted with their own beliefs or values.

4. The use of "dog whistles" in political communication is an example of how verbal communities can shape behaviour and communication. Dog whistles are coded messages that are intended to appeal to a specific group of individuals without being understood by others. For example, a political candidate might use a specific phrase or reference that is intended to appeal to a particular demographic, such as white supremacists or gun owners.

5. The use of medical jargon in the healthcare industry is an example of how verbal communities can shape behaviour and communication. Medical professionals often use technical language and jargon that is specific to their field, which can be confusing or intimidating to patients who are not familiar with that language. This can lead to misunderstandings or miscommunication between healthcare providers and patients, which can have negative consequences for patient outcomes.

These examples demonstrate how verbal communities can shape behaviour and communication in a wide range of ways, influencing everything from social hierarchies and political beliefs to healthcare outcomes and cognitive processes.

■ The role of verbal communities in behaviour modification
Verbal communities play an important role in behaviour modification by shaping the way that individuals communicate and interact with each other. These communities are made up of groups of people who share a common language or way of speaking, and they can have a significant impact on the way that individuals think, feel, and behave.

One way that verbal communities influence behaviour is through the use of social norms. Social norms are the unwritten rules and expectations that govern behaviour within a particular community. These norms are communicated through language and other forms of communication, and they can have a powerful impact on the way that individuals behave. For example, in some cultures, it is considered rude to interrupt someone while they are speaking. This norm is communicated through verbal and nonverbal cues, and individuals who violate this norm may be viewed negatively by others within the community. Verbal communities can also influence behaviour by providing individuals with a sense of identity and belonging. When individuals share a common language or way of speaking, they may feel a sense of connection and camaraderie with others within their community. This sense of identity and belonging can influence the way that individuals think and behave, and it can even influence their attitudes and beliefs. For example, individuals who are part of a religious community may adopt certain beliefs and behaviours that are consistent with the values of that community. Another way that verbal communities can influence behaviour is through the use of language as a tool for persuasion.

Language can be used to persuade individuals to adopt certain behaviours or attitudes, and verbal communities may use this tool to promote certain values or beliefs. For example, political campaigns may use language to persuade voters to support a particular candidate or issue. Finally, verbal communities can influence behaviour by shaping the way that individuals perceive and interpret the world around them. Language is a powerful tool for shaping perception, and individuals who are part of a particular verbal community may perceive the world differently than those who are not. For example, individuals who are part of a scientific community may perceive the world in terms of cause-and-effect relationships, while individuals who are part of a spiritual community may perceive the world in terms of divine intervention.

■ **Real-world examples of verbal communities in action**
- Children learning language from their parents or other caregivers
- Employees learning workplace language and jargon from colleagues and supervisors
- Social media influencers creating and spreading new slang or terminology
- Regional dialects or accents influencing language use and perception
- Religious groups using specific terminology and language to communicate and reinforce beliefs and values
- Sports teams developing their own language and terminology for plays and strategies
- Professional associations and organisations using industry-specific jargon and language
- Political groups using specific language and rhetoric to communicate and mobilise followers
- Online communities developing their own unique language and memes
- Families and friends develop inside jokes and slang that becomes part of their unique communication style.

Sometimes winning plays in football are communicated via secret signs amongst the verbal community of the football team.

Chapter 7: Relational Frame theory

■ Explanation of RFT

Relational Frame Theory (RFT) is a newer approach to behaviour analysis that is focused on the role of language, words, and thinking in human behaviour. At its core, RFT proposes that language is a relational behaviour that allows us to think about and manipulate concepts and ideas, and that this ability to think relationally shapes the way we perceive and interact with the world around us. Basically we have to use language to label things, objects, moments etc. The language acts as a frame of reference we can all understand when we hear it or read it. E.g. The red car was big. One of the key ideas in RFT is that our perception of the world is not just based on our direct experiences, but is also shaped by the frames through which we view the world. These frames can be thought of as the relational networks that we use to organise our experiences and make sense of the world around us. For example, we might have a frame that links the concept of "dog" to the concept of "barking," or a frame that links the concept of "money" to the concept of "power." These frames are not just passive structures that we use to understand the world, but are also actively constructed and reinforced through our use of language. For instance, when we use words like "good" or "bad" to describe something, we are not just making a value judgement, but are also reinforcing the relational network that links those words to the concepts we are describing. Because of the role that language plays in shaping our frames, RFT can be used to help us understand why people might have different perspectives on the same situation. Let's say two people are looking at a painting, one might see it as beautiful while the other sees it as ugly. RFT would suggest that these different perspectives are not just a matter of personal taste, but are also influenced by the frames that each person brings to the situation. Understanding the role of frames in shaping our perception of the world is an important part of behaviour modification, because it allows us to identify and modify the relational networks that underlie our behaviour. By helping individuals to recognize and change their frames, we can help them to develop more adaptive and functional behaviours, and to interact more effectively with the world around them.

■ **We are what we are because of Language**
Language is like a magic key that opens up our world. It's how we label and make sense of everything around us. Imagine you see a dog - in reality, it's just an animal with fur, four legs, and a wagging tail. But with language, it becomes a 'dog' to us, something we can talk about, remember, and interact with. Even more amazing, language helps us create a whole world inside our heads. That's why when you read a book or hear a story from a friend, you can 'see' the events in your mind as if you're watching a movie. This happens because of something called 'stimulus equivalence' - our brains connect words with images, emotions, and experiences, making them seem real to us. So, even though the world, the universe and reality exists independently of language, let's not forget before language what did humans call the sensation of anxiety? Or edible food? How did we explain advanced concepts? For us humans, language shapes our reality. It frames our environment and helps us interact with it and with each other. It's like we're all walking around with a personal movie projector in our heads, playing out the world in a way that makes sense to us, and all thanks to the power of language!

RFT proposes that humans have the ability to understand and manipulate relationships between things due to language. This ability is essential to human cognition and communication. This understanding allows us to create abstract concepts and to think about complex ideas. In RFT, relationships between words and concepts are referred to as relational frames. These frames are a kind of mental scaffolding that allow us to understand and use language in different ways. There are different types of relational frames, including opposition, comparison, hierarchy, and sequence. The significance of RFT lies in its ability to explain how people see differences in the world because of their frames. For example, two people might look at the same situation and see it in different ways because of the frames they use to interpret the situation. Understanding this concept can help us better communicate with others and bridge gaps in understanding. RFT also highlights the importance of context in shaping behaviour and communication. By understanding the relational frames that underpin our understanding of the world, we can better navigate social situations and communicate more effectively.

So we can surmise that relational framing refers to how we create meanings and associations between things in our minds based on our experiences and language. For example, when we see a cat, we might think of the word "furry" or "cute" based on past experiences with cats. These associations and meanings are created through our ability to relate one thing to another in our minds. When we engage in relational framing, we are essentially connecting the dots between different things and experiences. This process can happen both consciously and unconsciously, and can shape our thoughts, feelings, and behaviours in ways we may take for granted.

Also, linking back to verbal behaviour our verbal communities play a crucial role in shaping the language we use and thus, the way we frame our thoughts and experiences in RFT. By being a part of a particular community, we learn its unique set of verbal rules and ways of expressing ourselves, which can influence how we perceive and respond to the world around us. For example, imagine growing up in a community where it is common to use sarcasm as a form of humour. As a result, you may learn to interpret certain statements as sarcastic and respond with sarcasm in return. However, if you were raised in a different community where sarcasm was not used as frequently, you may not interpret those same statements as sarcastic and respond differently. I.e. You frame sarcasm from others as positive based on what you've learned from where you grew up. This concept is linked to the idea of behavioural agnosia in that our understanding and interpretation of the world is shaped by the environmental and cultural factors that surround us, which can impact our behaviour and thoughts in both positive and negative ways.

The vastness of language means we can apply words to things we see and experience!

■ **Overview of the six core processes of RFT**

RFT outlines six core processes that help us understand how language and thinking influence our behaviour. These six processes include:

1. Arbitrariness: This refers to the fact that the relationship between a word and its meaning is arbitrary, and it is established by social convention. For example, there is no inherent reason why the word "dog" should refer to the four-legged animal we call a dog.

2. Contextual framing: This refers to how words and concepts are interpreted differently depending on the context they are presented in. For example, the word "bear" might mean something very different in the context of a camping trip compared to a discussion on the stock market.

3. Transformation of stimulus functions: This refers to how words can transform the meaning of other words or stimuli. For example, hearing the word "fire" might evoke different responses depending on whether it is being used in the context of a campfire or a building on fire.

4. Entailment: This refers to the relationship between two stimuli where the presence of one stimulus implies the presence of the other. For example, the presence of a hammer implies the presence of a nail, and vice versa.

5. Relational framing: This refers to how words and concepts can be related to each other in different ways, such as opposition, comparison, or analogy. For example, the word "hot" might be related to "cold" through opposition, "bigger" might be related to "smaller" through comparison, and "life" might be related to "journey" through analogy.

6. Perspective-taking: This refers to the ability to take the perspective of others and to understand that their perspective may differ from our own. This is important for developing empathy and effective communication.

Understanding these six core processes can help us understand how language and thinking influence our behaviour and how we perceive the world around us. By becoming aware of these processes, we can develop more effective communication skills and improve our ability to understand and relate to others.

■ VB and RFT?

Verbal behaviour and RFT are both concepts that can seem daunting and confusing at first glance, especially for those who are not familiar with the field of behaviour analysis. However, at their core, they are both focused on the ways in which humans communicate and make sense of the world around them. Verbal behaviour refers to the use of language to communicate with others, including everything from spoken words to written texts and gestures. It is an essential part of how humans interact with one another, and is a key tool for shaping and modifying behaviour. On the other hand, RFT is a theory that helps explain how humans use language to create and understand relationships between different concepts and ideas. In essence, RFT is concerned with the way that we use language to make sense of the world around us, and how those relationships and understandings can influence our behaviour. While the two concepts may seem complex, they are both based on the idea that language is a powerful tool for shaping behaviour and understanding the world. By exploring the ways in which we use language to communicate and form relationships between different ideas and concepts, we can better understand how behaviour is influenced and modified over time. So, whether you are interested in the way that humans communicate or the ways in which behaviour can be shaped and modified, understanding the concepts of verbal behaviour and RFT is a critical first step towards gaining a deeper understanding of the world around us.

■ **Examples of how RFT can be used to modify behaviour**

1. Overcoming prejudice: RFT can help individuals to recognize and challenge their own prejudicial attitudes and beliefs by highlighting the arbitrary nature of language-based categories. For example, someone might realise that the label "Black" or "White" is not an inherent quality of a person, but rather a social construct that has been arbitrarily assigned based on historical and cultural factors. By recognizing this, they can work to reduce their prejudicial thoughts and behaviours.

2. Improving interpersonal relationships: RFT can help individuals to communicate more effectively with others by teaching them to be more aware of the relational frames that are at play in their interactions. For example, someone might learn to recognize when they are engaging in "I-Thou" thinking (i.e., viewing the other person as a unique individual with their own thoughts and feelings) versus "I-It" thinking (i.e., viewing the other person as an object to be manipulated or controlled). By becoming more aware of these relational frames, they can work to improve their interpersonal relationships.

3. Addressing anxiety and depression: RFT can help individuals to overcome negative thinking patterns that contribute to anxiety and depression. For example, someone might learn to recognize when they are engaging in "cognitive fusion" (i.e., being overly attached to their thoughts and feelings) versus "cognitive defusion" (i.e., recognizing that their thoughts and feelings are just passing events in the mind). By practising cognitive defusion, they can learn to detach from negative thoughts and feelings and reduce their impact on their mood and behaviour.

4. Enhancing creativity: RFT can help individuals to break out of habitual thinking patterns and generate more creative ideas by encouraging them to think in new and different ways. For example, someone might learn to use "deictic framing" (i.e., shifting their perspective to see things from a different point of view) to come up with novel solutions to problems or to see things in a new light.

5. Increasing mindfulness: RFT can help individuals to become more aware of their present-moment experience and to detach from unhelpful thoughts and feelings. For example, someone might learn to practise "self-as-context" (i.e., recognizing that their thoughts and feelings are just passing events in the mind and that they are the context in which those events occur) to become more mindful and present in their daily life.

6. Promoting empathy and compassion: RFT can help individuals to develop more empathy and compassion for others by teaching them to recognize the common humanity that underlies all human experience. For example, someone might learn to recognize the "relational frame of similarity" (i.e., recognizing that they share common experiences and emotions with others, even if they appear different on the surface) to cultivate greater empathy and compassion towards others.

These are just a few examples of how RFT can be applied to different social issues, but there are many more. The key is to recognize the power of language and relational frames in shaping our experience of the world and to use this understanding to modify behaviour in positive ways.

Diversity and inclusion in our relational framing can open our world to new experiences

Chapter 8: Putting it into Practice

■ **Using behavioural Principles in Everyday Life.**

In this final chapter, we will summarise all the important points covered in the book so far and show how they can be applied to the reader's daily life. We have covered a lot of ground, so let's take a moment to recap. We started with the concept of behavioural agnosia, which refers to people's lack of awareness of the behavioural principles that influence their daily lives. This blindness can lead to problems in our relationships, work, and personal lives. We then shared some examples of the kinds of people who may have experienced Behavioural Agnosia. Here are some potential examples of the unacknowledged factors behind them:

Individuals who struggle with academic performance, particularly those who may have a difficult time connecting with and understanding the material presented to them. By identifying the antecedents and SDs of their difficulties, such as distractions - social (such as a mate making faces) or mental (via verbal self-talk), they can work to modify their behaviour and improve their academic performance.

Those who commit crimes may experience behavioural agnosia as they fail to recognize the discriminative stimuli (such as peer pressure or negative emotions) that trigger their criminal behaviour. Also the words they use like in UK Drill culture a "Lick" is someone to rob. Seeing them not as individual humans, but as a potential means to make money .

Individuals with sudden and unexplained mental health issues may experience behavioural agnosia if they are unable to identify the antecedents (such as stressful events or triggers) that lead to their symptoms. By identifying these antecedents and learning coping strategies, they can work to manage their symptoms.

People with varying personalities may experience behavioural agnosia if they are unable to recognize the environmental factors (such as social cues or expectations) that trigger their different personas. By identifying these factors and practising self-awareness, they can work to better understand and manage their behaviour in different situations.

Stressed single parents who may struggle to manage their own emotions and the behaviour of their children. Negative self-talk or beliefs about one's ability to parent effectively, which can lead to feelings of overwhelm or frustration. Difficulty in recognizing or responding to the antecedents or consequences of children's behaviour, which can contribute to a cycle of negative interactions between parent and child.

Friends, family members, or colleagues who exhibit sudden changes in mood or behaviour, particularly if these changes are uncharacteristic or difficult to understand. By using the principles of ABA to identify the antecedents and consequences of their behaviour, they can work to understand the reasons behind the changes and modify their behaviour accordingly.

■ We then explored the basics of behaviourism, including operant and respondent conditioning, reinforcement, punishment, and extinction. These principles play a crucial role in shaping our behaviour.

■ The ABCs of behaviour, which stand for Antecedent, behaviour, and Consequence, were also introduced. This framework helps us understand the factors that influence our behaviour and can be used to modify it.

■ Motivating operations, such as establishing and abolishing operations, were discussed. These operations can have a significant impact on our behaviour and are important to understand when attempting to modify it.

■ We then delved into the concept of discriminative stimuli and how they can influence behaviour. We explored how different types of stimuli can shape our behaviour in various ways.

■ Verbal behaviour and verbal communities were also discussed. We explored how language can shape our behaviour and how our communities can influence our use of language.

■ Finally, we introduced the concept of RFT and how it can be used to modify behaviour. We discussed the six core processes of RFT and explored how they can be applied to social issues.

Now that we have covered these topics, let's take a moment to reflect on how they can be applied to our daily lives. Here are some ways in which we can use this knowledge to improve our behaviour:

■ We can use the ABCs of behaviour to identify the factors that influence our behaviour and modify them as needed.

■ By understanding motivating operations, we can create an environment that encourages the behaviours we want to see and discourages those we don't.

■ Knowing about discriminative stimuli can help us understand why we behave the way we do in certain situations and how we can modify our behaviour.

■ We can use our knowledge of verbal behaviour to improve our communication skills and build better relationships with others.

■ By being aware of the communities we belong to, we can recognize how they influence our behaviour and take steps to modify it if necessary.

■ Finally, by understanding RFT and its core processes, we can approach social issues with a new perspective and work towards behaviour change on a larger scale.

In conclusion, the principles of behaviourism and the concepts we have explored in this book can have a significant impact on our daily lives. By understanding these principles and applying them in our personal and professional lives, we can improve our relationships, work, and overall well-being.

Now you're free to reinterpret the world without Behavioural Agnosia!

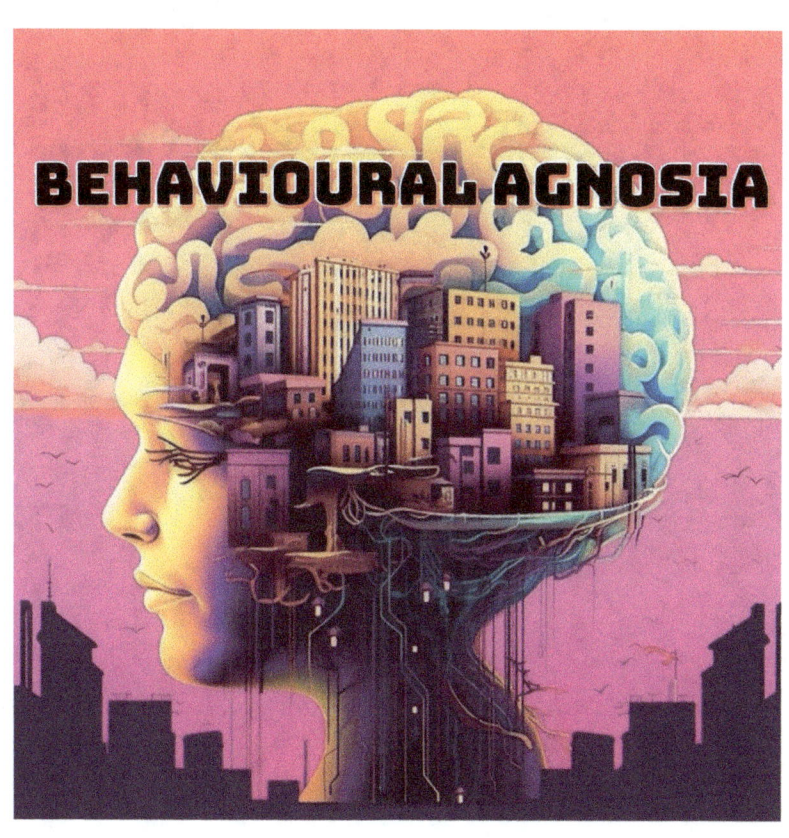

With special thanks to;

My Partner

Our Beautiful Baby Girl

My amazing mother

My astute older brother

My brave younger brother

.

My hard working brother

My talented youngest brother

And our feisty younger sister

.